220
Advances in Polymer Science

Advances in Polymer Science
Recently Published and Forthcoming Volumes

Self-Assembled Nanomaterials II

Nanotubes

Volume Editor: Toshimi Shimizu

With contributions by

T. Aida · T. Fukushima · G. Liu · M. Numata
S. Shinkai · M. Steinhart · T. Yamamoto

 Springer

The series *Advances in Polymer Science* presents critical reviews of the present and future trends in polymer and biopolymer science including chemistry, physical chemistry, physics and material science. It is adressed to all scientists at universities and in industry who wish to keep abreast of advances in the topics covered.
As a rule, contributions are specially commissioned. The editors and publishers will, however, always be pleased to receive suggestions and supplementary information. Papers are accepted for *Advances in Polymer Science* in English.
In references *Advances in Polymer Science* is abbreviated *Adv Polym Sci* and is cited as a journal.

Springer WWW home page: springer.com
Visit the APS content at springerlink.com

ISBN 978-3-540-85104-2   e-ISBN 978-3-540-85105-9
DOI 10.1007/978-3-540-85105-9

Advances in Polymer Science ISSN 0065-3195

Library of Congress Control Number: 2008933504

Cover design: WMXDesign GmbH, Heidelberg
Typesetting and Production: le-tex publishing services oHG, Leipzig

Printed on acid-free paper

9 8 7 6 5 4 3 2 1 0

springer.com

Advances in Polymer Science
Also Available Electronically

For all customers who have a standing order to Advances in Polymer Science, we offer the electronic version via SpringerLink free of charge. Please contact your librarian who can receive a password or free access to the full articles by registering at:

springerlink.com

If you do not have a subscription, you can still view the tables of contents of the volumes and the abstract of each article by going to the SpringerLink Homepage, clicking on "Browse by Online Libraries", then "Chemical Sciences", and finally choose Advances in Polymer Science.

You will find information about the

- Editorial Board
- Aims and Scope
- Instructions for Authors
- Sample Contribution

at springer.com using the search function.

Color figures are published in full color within the electronic version on SpringerLink.

Preface

Nanotechnology is the creation of useful materials, devices, and systems through the control of matter on the nanometer-length scale. This takes place at the scale of atoms, molecules, and supramolecular structures. In the world of chemistry, the rational design of molecular structures and optimized control of self-assembly conditions have enabled us to control the resultant self-assembled morphologies having 1 to 100-nm dimensions with single-nanometer precision. This current research trend applying the bottom-up approach to molecules remarkably contrasts with the top-down approach in nanotechnology, in which electronic devices are miniaturizing to smaller than 30 nm. However, even engineers working with state-of-the-art computer technology state that maintaining the rate of improvement based on Moore's law will be the most difficult challenge in the next decade.

On the other hand, the excellent properties and intelligent functions of a variety of natural materials have inspired polymer and organic chemists to tailor their synthetic organic alternatives by extracting the essential structural elements. In particular, one-dimensional structures in nature with sophisticated hierarchy, such as myelinated axons in neurons, tendon, protein tubes of tubulin, and spider webs, provide intriguing examples of integrated functions and properties.

Against this background, supramolecular self-assembly of one-dimensional architectures like fibers and tubes from amphiphilic molecules, bio-related molecules, and properly designed self-assembling polymer molecules has attracted rapidly growing interest. The intrinsic properties of organic molecules such as the diversity of structures, facile implementation of functionality, and the aggregation property, provide infinite possibilities for the development of new and interesting advanced materials in the near future. The morphologically variable characteristics of supramolecular assemblies can also function as pre-organized templates to synthesize one-dimensional hybrid nanocomposites. The obtained one-dimensional organic–inorganic, organic–bio, or organic–metal hybrid materials are potentially applicable to sensor/actuator arrays, nanowires, and opto-electric devices.

The present volumes on Self-Assembled Nanofibers (Volume 219) and Nanotubes (Volume 220) provide an overview on those aspects within eight chapters. Different points of view are reflected, featuring interesting aspects related to (a) the self-assembly of supramolecular nanofibers comprising of organic, polymeric, inorganic and biomolecules (N. Kimizuka, in Volume 219, Chapter 1), (b) controlled self-assembly of artificial peptides and peptidomimetics into nanofiber architectures (N. Higashi, T. Koga, in Volume 219, Chapter 2), (c) self-assembled nanostructures from amphiphilic rod molecules (B.-K. Cho, H.-J. Kim, Y.-W. Chung, B.-I. Lee, M. Lee, in Volume 219, Chapter 3), (d) the production of functional self-assembled nanofibers by electrospinning (A. Greiner, J. H. Wendorff, in Volume 219, Chapter 4), (e) the synthesis of tailored π-electronic organic nanotubes and nanocoils (T. Yamamoto, T. Fukushima, T. Aida, in Volume 220, Chapter 1), (f) preparation and fundamental aspects of nanotubes self-assembled from block copolymers (G. Liu, in Volume 220, Chapter 2), (g) β-1,3-glucan that can act as unique natural nanotubes and incorporate conjugated polymers or molecular assemblies (M. Numata, S. Shinkai, in Volume 220, Chapter 3), and (h) the fabrication of self-assembled polymer nanotubes involving the use of a nanoporous hard template (M. Steinhart, in Volume 220, Chapter 4). A variety of nanofibers and nanotubes with well-defined morphologies and dimensions are discussed in terms of self-assembly of molecular and polymer building blocks in bulk solution or confined geometry like nanopores.

Current materials and manufacturing technologies strongly require technological advances for reducing environmental load combined with energy and resource savings in production. In order to develop such technologies for the development of a sustainable society, research on materials production based on the self-assembly technique is of great interest. Hopefully, these volumes will be beneficial to readers involved with self-organization in the field of bottom-up nanotechnology as well as those concerned with industrial fiber processing.

Tsukuba, June 2008   Toshimi Shimizu

Contents

Contents of Volume 219

Adv Polym Sci (2008) 220: 1–27
DOI 10.1007/12_2008_171
© Springer-Verlag Berlin Heidelberg
Published online: 15 August 2008

Self-Assembled Nanotubes and Nanocoils from π-Conjugated Building Blocks

Takuya Yamamoto[1] · Takanori Fukushima[2,3] (✉) · Takuzo Aida[1,2,3] (✉)

[1] ERATO-SORST Nanospace Project, Japan Science and Technology Agency,
National Museum of Emerging Science and Innovation, 2-41 Aomi, Koto-ku,
135-0064 Tokyo, Japan

[2] Advanced Science Institute, RIKEN, 2-1 Hirosawa Wako, 351-0198 Saitama, Japan
fukushima@riken.jp

[3] Department of Chemistry and Biotechnology, School of Engineering,
The University of Tokyo, 7-3-1 Hongo, Bunkyo-ku, 113-8656 Tokyo, Japan
aida@macro.t.u-tokyo.ac.jp

Abstract This review article describes recent studies on the self-assembly and co-assembly of π-conjugated molecules into nanotubes and nanocoils. Such π-conjugated molecules include phenylenes, phenylene vinylenes, phenylene ethynylenes, porphyrins, phthalocyanines, and polycyclic aromatic hydrocarbons, which are properly modified with hydrophilic and/or hydrophobic side chains for cooperative interactions. Not only nanocoils but also most reported nanotubes possess a helical chirality. These assembling events possibly show chiral amplification, where one-handedness can be realized even from stereochemically impure components (majority rule). Combination of components leading to non-tubular assemblies with properly chosen chiral components may give rise to nanotubes or nanocoils with one-handed helical chirality (sergeants-and-soldiers effect). Covalent modification of assembled components can enhance physical robustness against heating and solvolysis.

Keywords π-Conjugation · Chirality · Nanocoil · Nanotube · Self-assembly

1
Introduction

Since the discovery of carbon nanotubes [1,2], π-electronic tubular nano-objects have attracted great interest, and they have been explored as application-oriented materials due to unique characteristics originating from this particular morphology. While many carbon nanotubes-based functional materials have been developed by relying on their remarkable properties, such as excellent electrical conductivity and mechanical strength [3], the selective production and isolation of carbon nanotubes with certain structural and electronic properties remain unsolved [4]. Meanwhile, the superb aspects of carbon nanotubes have motivated chemists to tailor their organic alternatives by extracting the essential structural element, i.e., extended π-conjugation. Such π-electronic organic nanotubes potentially have a large design flexibility in functionalization and allow precise dimensional control through elaboration of molecular building blocks, thereby giving an opportunity for the fabrication of low-dimensional soft materials with a wide variety of electronic and optoelectronic properties. However, for the construction of such a hollow cylindrical morphology, one has to address much more rigorous requisites for the molecular orientation than other molecular assemblies including fibers and vesicles. In other words, implementation of highly sophisticated self-assembling programs is needed.

This review article focuses attention on recent progress in the synthesis of tailored π-electronic organic nanotubes and nanocoils. A coiled structure is a loosened form of a tube consisting of a rolled-up tape [5,6]. Although many examples of twisted ribbons formed from π-electronic molecules have been reported [7–9], they are not included in this review article, because twisted ribbons do not have the structural element to generate nanotubes, while nanocoils are en route to become tubules. Molecular building blocks with extended π-conjugation, featured in this article, include oligomers of phenylenes, phenylene vinylenes, and phenylene ethynylenes, organic dyes such as porphyrins and phthalocyanines, and polycyclic aromatic hydrocarbons (PAHs), such as hexa-peri-hexabenzocoronenes. These compounds self-assemble via π-stacking as the primary driving force, which is properly modified by other complementary forces to achieve molecular geometries needed for the tubular morphology. Resulting nanotubes, consisting of ordered π-stacked arrays, can allow directional transports of energy and charge carriers, and are potent components for organic electronic and optoelectronic devices [10].

2
π-Conjugated Linear Oligomers

Linear oligomers including oligo(p-phenylenes), oligo(p-phenylene vinylenes), and oligo(p-phenylene ethynylenes) are one of the most extensively studied π-conjugated building blocks for self-assembly [11]. The rigid π-conjugated backbones, when functionalized with hydrophilic or hydrophobic side chains, have been reported to form low-dimensional nanostructured assemblies via cooperative π-stacking and side chain interactions. While most of these examples give nanofibers [10, 11], only a few compounds are reported to self-assemble into nanotubes.

2.1
Nanotubes from Oligo(p-Phenylenes)

Self-assembly of amphiphilic oligo(p-phenylene) derivatives has been extensively studied by Lee et al. Macrocyclic compound **1** (Fig. 1a) composed of a rigid hexa(p-phenylene) unit in conjunction with a flexible oligoether chain containing eighteen oxyethylene units and a chiral ether unit, originating from 1,2-epoxypropane, at each of the two termini self-assembles into nanotubes in water [12]. According to TEM, the nanotubes possess a diameter of 20 nm and a wall thickness of 3 nm (Fig. 1c). Upon being stained with uranyl acetate, the nanotubes show a left-handed helical stripe pattern with a regular pitch of 4.7 nm. Furthermore, an aqueous solution of self-assembled **1** is active in circular dichroism (CD). Based on these observations coupled with results of a small angle X-ray scattering analysis, the authors claim that a tape-like structure consisting of a π-stacked bilayer of **1** is initially formed and then rolled up with a preferred handedness to give the tubular structure (Fig. 1b).

Another type of rod–coil block molecule **2** (Fig. 2a) consists of a rigid oligophenylene-based macrocycle appended with two oligo(ethylene oxide) dendrons that forms cylindrical micelles in water by stacking directly on top of each other [13]. Hence, the assembling manner of **2** is quite different from that of **1**. Dynamic light scattering analysis of an aqueous solution of **2** suggests the presence of cylindrical micelles. The average diameter of the micelles, as evaluated by TEM microscopy, is 10 nm (Fig. 2b), which is consistent with the dimensions of **2** estimated by its CPK model.

The nanotube of **2** is able to solubilize single-walled carbon nanotubes (SWNTs) in water. On solubilization of SWNTs, the organic nanotubes do not change their dimensions, as confirmed by TEM (Fig. 2c), but show significant fluorescence quenching. Based on these observations, the authors suggest that tubularly assembled **2** encapsulates SWNTs into its hydrophobic hollow space. In contrast, compound **2** is unable to solubilize SWNTs in THF, where **2** is molecularly dispersed. Therefore, not only the ring-shape of

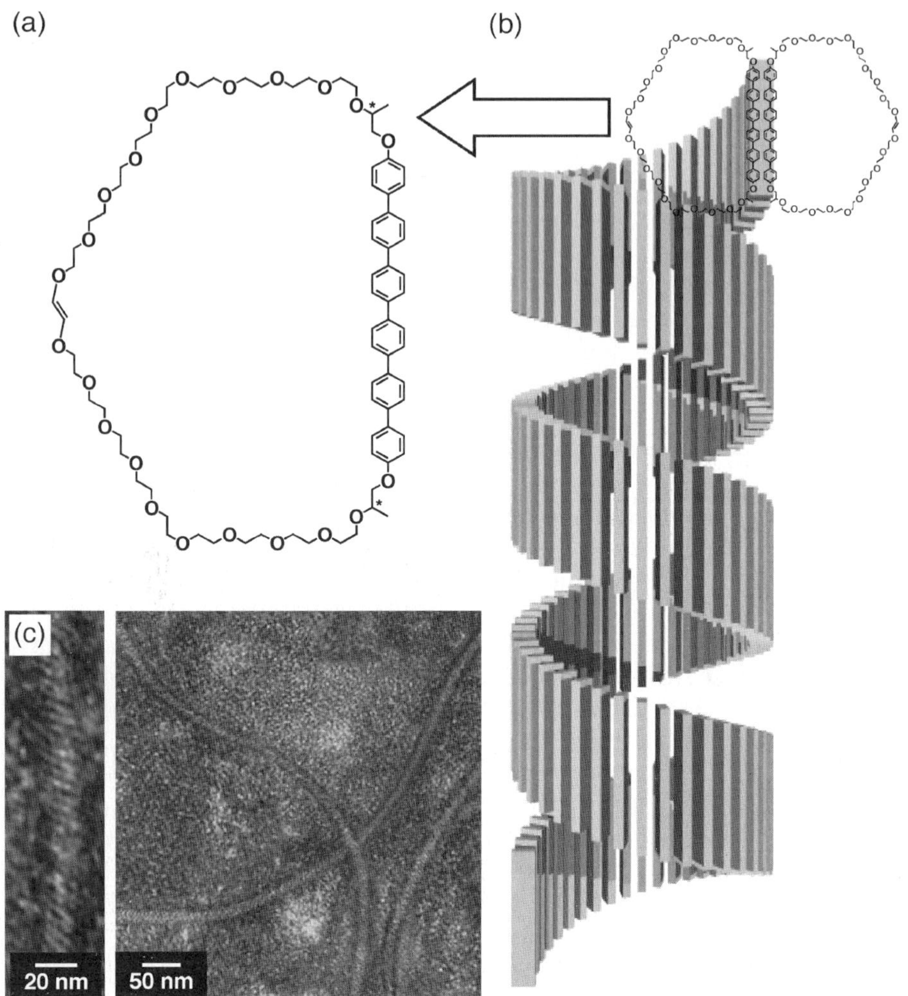

Fig. 1 Molecular structure of **1** (**a**). Proposed structure of the nanotube of self-assembled **1** (**b**). TEM micrograph of the nanotubes of self-assembled **1**. The TEM micrographs were provided courtesy of Prof. Myongsoo Lee of Yonsei University (**c**)

2 but also its laterally assembled structure is important for the solubilization of SWNTs.

2.2
Nanocoils from Oligo(*p*-Phenylene vinylenes)

Ajayaghosh et al. have reported nanocoiled assemblies of a short-chain oligo(*p*-phenylene vinylene) derivative **3** (Fig. 3a) in dodecane [14]. It is noteworthy that formation of the coiled assembly requires the presence of

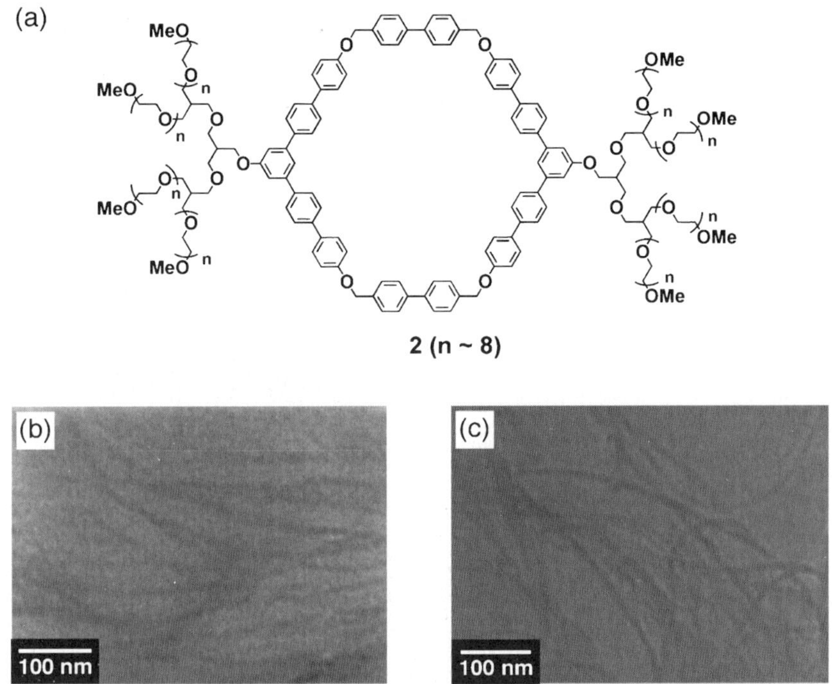

Fig. 2 Molecular structure of **2** (**a**). TEM micrographs of the nanotubes of self-assembled **2** (**b**) and those formed in the presence of carbon nanotubes (**c**). The TEM micrographs were provided courtesy of Prof. Myongsoo Lee of Yonsei University

a stereogenic center in the side chains of **3** (**3b**), while the assembly of its achiral version (**3a**) results in the formation of nanofibers [15]. The self-assembled nanocoils from (*S*)-**3b** show bisignate CD signals at the absorption bands for the π–π* transition. Atomic force microscopy (AFM) allows for the visualization of left-handed nanocoils that are several micrometers long and 100 nm wide with a helical pitch of roughly 100 nm (Fig. 3b). When fiber-forming achiral **3a** is allowed to co-assemble with coil-forming chiral **3b**, nanocoils form exclusively. This phenomenon may be referred to as the sergeants-and-soldiers effect [16, 17], where a polymer or assembly of an achiral "soldier" component adopts a prevailing one-handed helical chirality when it accommodates a chiral "sergeant" component. However, unlike other examples, the handedness of the nanocoils changes as a function of the sergeant/soldier composition. When the mole fraction of (*S*)-**3b** is higher than 60 mol %, the most nanocoils are left-handed as in the case of the homoassembly of (*S*)-**3b**. In sharp contrast, when the mole fraction of (*S*)-**3b** is lower than 22 mol %, mistranslation of the sergeant's chirality occurs, resulting in the preferential formation of right-handed nanocoils (Fig. 3c).

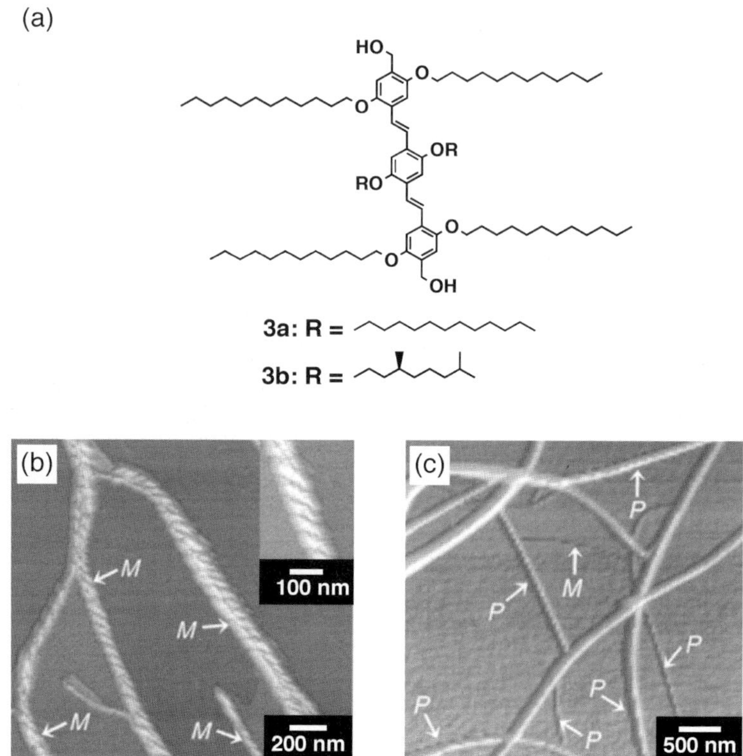

Fig. 3 Molecular structures of **3a** and **3b** (**a**). AFM images of the nanocoils from **3b** by self-assembly (**b**) and co-assembly of **3a** with 9 mol % of **3b** (**c**). *M*: left-handed. *P*: right-handed. The AFM images were provided courtesy of Prof. Ayyappanpillai Ajayaghosh of National Institute for Interdisciplinary Science and Technology

2.3
Nanotubes from Oligo(*p*-Phenylene ethynylenes)

Ajayaghosh et al. have also reported the successful formation of nanotubes via co-assembly of short-chain oligo(*p*-phenylene ethynylene) derivatives **4a** and **4b** (Fig. 4a), which are structurally analogous to **3** [18]. However, the homo-assembling and co-assembling behaviors are essentially different from those of **3**. Achiral **4a** self-assembles in hydrocarbon solvents to form a vesicular structure with a diameter of approximately 100 nm (Fig. 4c) [19]. In contrast, chiral **4b** does not aggregate under similar conditions. Of interest, co-assembly of **4a** with 25 mol % of **4b** leads to the quantitative formation of helical nanotubes (Fig. 4c), whose dimensions, as determined by AFM, are 90 nm or larger in width and 140 nm in helical pitch. TEM microscopy reveals that the inner diameter of the nanotubes ranges from 55

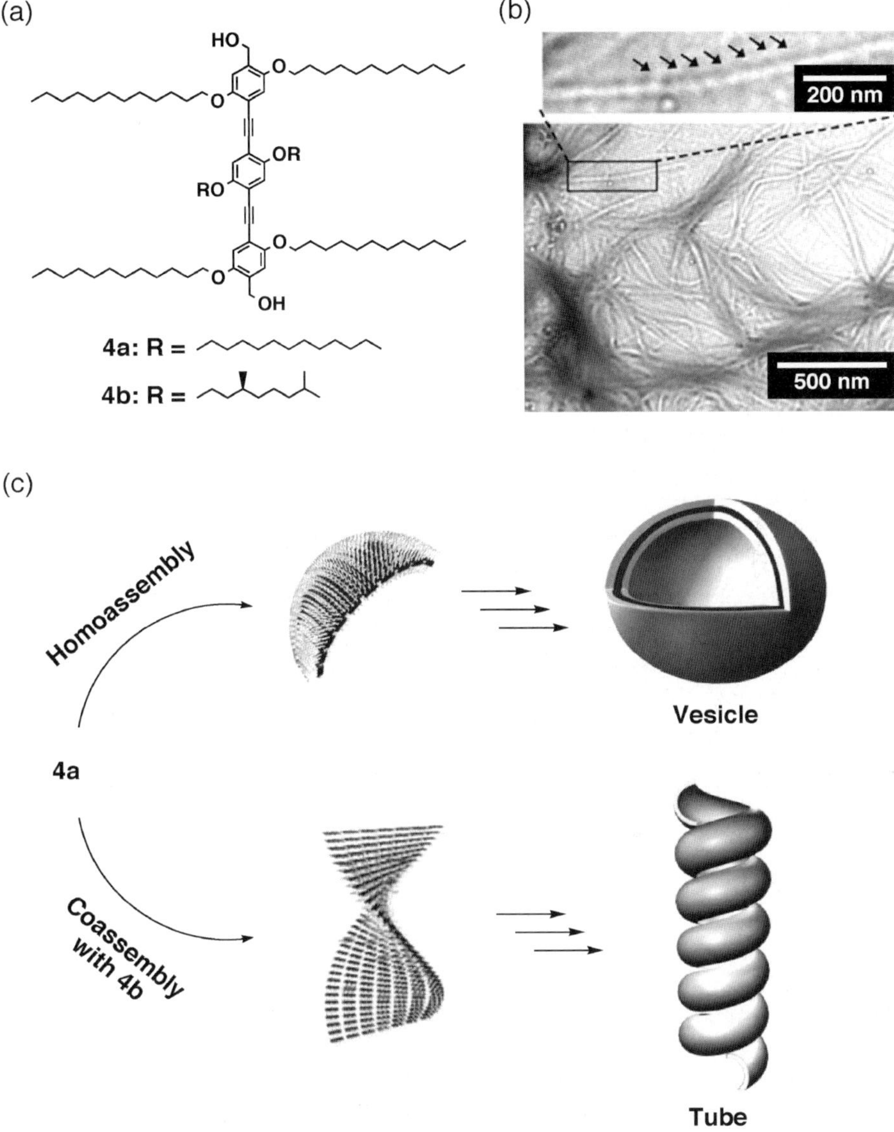

Fig. 4 Molecular structures of **4a** and **4b** (**a**). TEM micrograph of the nanotubes of co-assembled **4a** with 25 mol % of **4b** (**b**). Proposed mechanisms for the nanotubular and vesicular co-assemblies of **4a** with **4b** (**c**). The TEM micrographs and illustrations of molecular arrays, vesicle, and tube were provided courtesy of Prof. Ayyappanpillai Ajayaghosh of National Institute for Interdisciplinary Science and Technology

to 90 nm (Fig. 4b). It should be noted however that the selective formation of the nanotubes takes place only in a limited range of the mole fraction of **4b**. For example, when 8 mol % of **4b** are used, both vesicles and nanotubes form. In contrast, attempted co-assembly with 50 mol % of **4b** gives rise to irregular aggregates. Accordingly, CD spectral profiles of the co-assembling system are composition-dependent. When the mole fraction of **4b** is in a range of 0–30 mol %, the CD intensity increases in proportion to the mole fraction of **4b**. However, further increase in the mole fraction of **4b** results in a decrease of the CD intensity. Thus, only the nanotubes are CD active.

3
Porphyrins and Phthalocyanines

Since porphyrin and phthalocyanine derivatives possess excellent electronic and photophysical properties, they have been extensively studied as functional components for a wide variety of crystalline, liquid crystalline, and polymeric materials. Increasing attention has also been paid to well-defined nanostructured assemblies of such organic dyes, and indeed, many examples of fibers, tapes, and vesicles have been reported [20]. However, only two examples are known for the formation of nanotubes and nanocoils.

Shelnutt et al. have reported the formation of nanotubes by co-assembly of positively and negatively charged porphyrin derivatives [21]. When porphyrins **5** (Fig. 5a) and **6a** (Fig. 5b), appended with sulfonate and 4-pyridyl groups, respectively, are allowed to co-assemble in water, nanotubes with 50–70 nm in diameter and approximately 20 nm in wall thickness are yielded, as observed by TEM microscopy (Fig. 5c). The use of **6b** having 3-pyridyl groups, instead of **6a**, for the co-assembly with **5** leads to the formation of nanotubes with a smaller diameter (35 nm) and a larger wall thickness. On the other hand, no tubular object results when **6c** with 2-pyridyl groups is used for the co-assembly. It is noteworthy that the central metal ion of the pyridylporphyrin affects the nanotube formation. While similar nanotubes form when Sn^{4+} in **6a** is replaced with other six-coordinate metal ions such as Fe^{3+}, Co^{3+}, TiO^{2+}, and VO^{2+}, the use of Cu^{2+}, which lacks the axial coordination capability, does not give a tubular structure. It should also be noted that the successful co-assembly into nanotubes takes place only at pH 2. Even a slight change of the pH value (e.g., pH 1 and 3) results in failure of the nanotube formation, suggesting that the degree of protonation of the sulfonate group of **5** plays a crucial role. As evaluated by electronic absorption and energy-dispersive X-ray spectroscopy, the mole ratio of **5** to **6a** in the nanotubes is 2.0–2.5, which may reflect the charge balance between **5** and **6a** at pH 2. Interestingly, upon exposure to an incandescent light, the co-assembled nanotubes transform into a non-hollow cylindrical structure. This morphological change is reversible as the tubular structure restores when the cylindrical rods are allowed to stand in the dark. The authors imply that a pho-

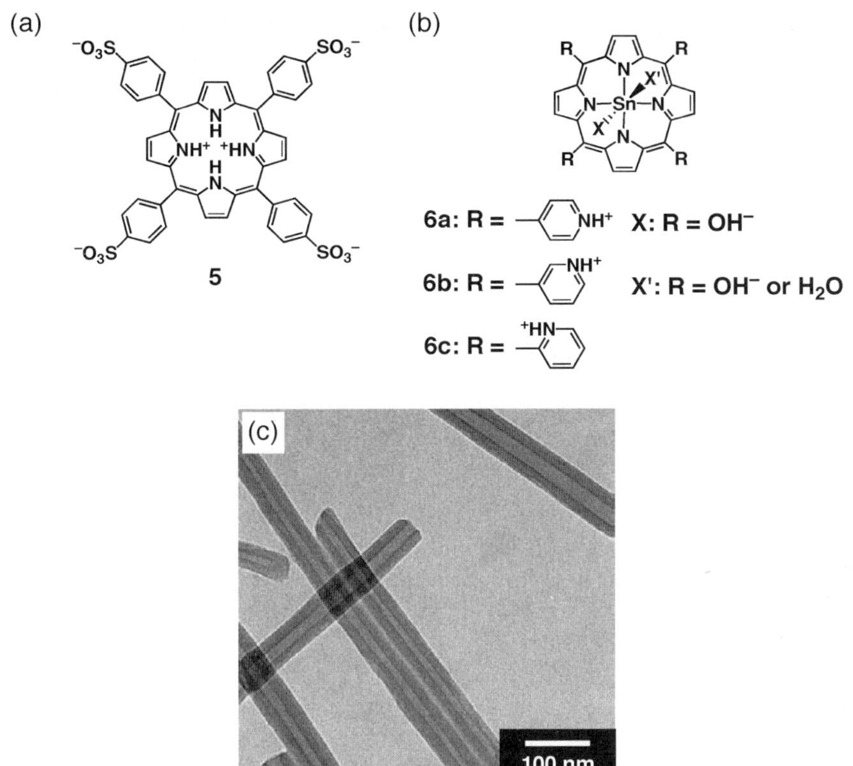

Fig. 5 Molecular structures **5** (**a**) and **6a–c** (**b**). TEM micrograph of the nanotubes of co-assembled **5** with **6a**. The TEM micrograph was provided courtesy of Prof. John A. Shelnutt of Sandia National Laboratories (**c**)

toinduced electron transfer from **5** to **6a**, which alters the charge balance, so that the tubular assembly becomes dynamic, is a possible mechanism.

Nolte et al. have reported that phthalocyanine (Pc) derivative **7** carrying four tetrathiafulvalene (TTF) units via a crown ether spacer (Fig. 6) self-assembles into nanocoils [22]. When dioxane is added to a $CHCl_3$ solution of **7** (12 mg ml^{-1}), gelation takes place. As observed by TEM microscopy, the gel after being dried contains several micrometer-long thin tapes with approximately 20 nm in width, some of which roll up to form nanocoils. Based on model studies, the authors suggest that the interactions operative in the TTF–TTF and TTF–Pc units are responsible for the formation of the tapes, where the Pc core adopts an offset π-stacking geometry.

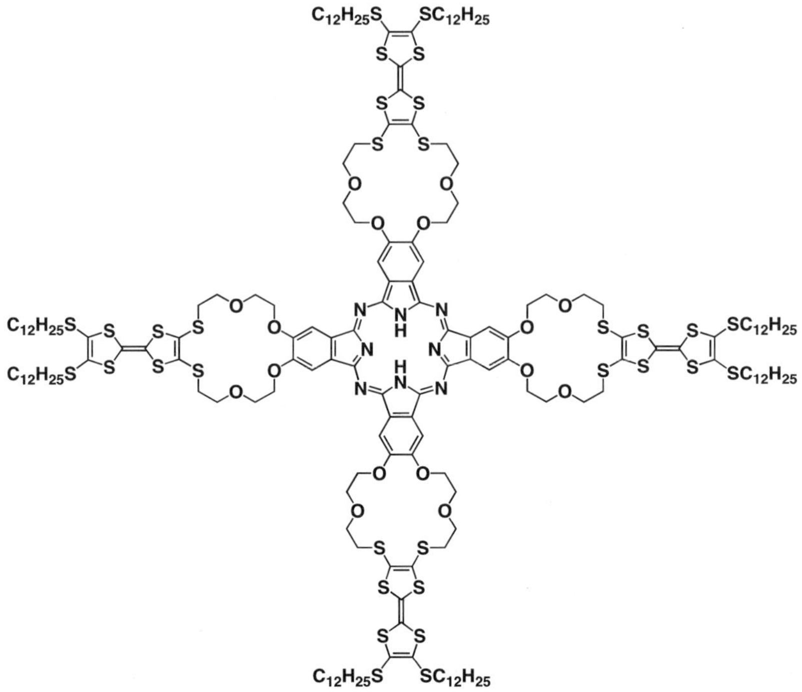

Fig. 6 Molecular structure of 7

4
Polycyclic Aromatic Hydrocarbons

4.1
Hexa-peri-hexabenzocoronenes (HBCs)

Graphene sheet (Fig. 7b) is the structural element of carbon nanotubes (Fig. 7a). While such an infinite carbon sheet is unattainable by organic synthesis, Müllen et al. have established the synthesis of a family of its small fragments [23]. Hexa-peri-hexabenzocoronene (HBC), which consists of 13 fused benzene rings (Fig. 7c) is a representative of such "molecular graphenes". The first synthesis of HBC was reported by Clar et al. in 1959 [24], which was followed by pioneering works of Müllen et al. on self-assembling HBCs with paraffinic side chains. In particular, the discovery of the formation of discotic liquid crystals has opened a new potential of this π-conjugated building block for electronic and optoelectronic materials [25–28]. More recently, by introducing a molecular design concept with amphiphilicity, Aida, Fukushima, et al. have developed a new class of HBCs that can self-assemble into well-

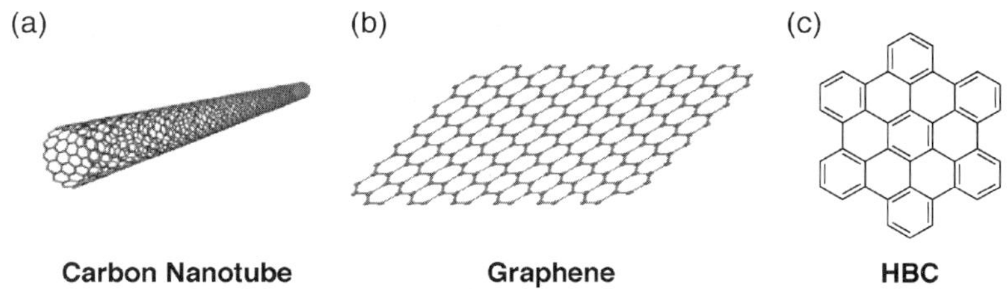

Fig. 7 Schematic representations of carbon nanotube (**a**), graphene (**b**), and hexa-peri-hexabenzocoronene (HBC) (**c**)

defined nanotubes [29]. The following section focuses attention on the design and self-assembling behaviors of a variety of HBC nanotubes.

4.1.1
Self-Assembled HBC Nanotubes

The first nanotubular assembly from HBC was realized by Gemini-shaped amphiphilic derivative **8a** (Fig. 8a), which carries two triethylene glycol (TEG) chains on one side and two dodecyl (C_{12}) chains on the other [29]. When a THF suspension of **8a** (1 mg ml^{-1}) is once heated at 50 °C, and the resulting homogeneous solution is allowed to cool to room temperature, self-assembly of **8a** takes place quantitatively to form nanotubes (Fig. 8d). The nanotubular assembly is yellow-colored with red-shifted absorption bands at 426 and 459 nm (Fig. 8b), and isolable without disruption by filtration. SEM microscopy clearly displays that the nanotube ends are open (Fig. 8c). As shown by a TEM micrograph in Fig. 8e, the nanotubes have a very high aspect ratio (> 1000) and a uniform diameter of 20 nm with a wall-thickness of 3 nm. By means of electron diffraction analysis, the HBC units are π-stacked with a plane-to-plane separation of 3.6 Å, which is comparable to that of the (002) diffraction of graphite (3.35 Å), indicating that the tubular wall consists of a great number of π-stacked HBC units. Moreover, infrared spectroscopy shows CH_2 stretching vibrations at 2917 (ν_{anti}) and 2848 (ν_{sym}) cm^{-1}, characteristic of paraffinic chains with a stretched conformation. Thus, the dodecyl side chains most likely interdigitate with one another to form a bilayer structure. Interestingly, when **8a** is allowed to self-assemble in a mixture of THF and water (8/2 v/v), a coiled structure (Fig. 8f and g) results along with the nanotubes. Thus, the nanotubes are likely formed by rolling-up of a two-dimensional pseudo-graphite tape composed of bilaterally coupled columns of π-stacked **8a** (Fig. 9). Here, the interdigitated dodecyl chains hold the bilayer structure, while the hydrophilic TEG chains, located on both sides of the bilayer tape, may suppress the formation of multi-lamellar structures unfavorable for the tube formation.

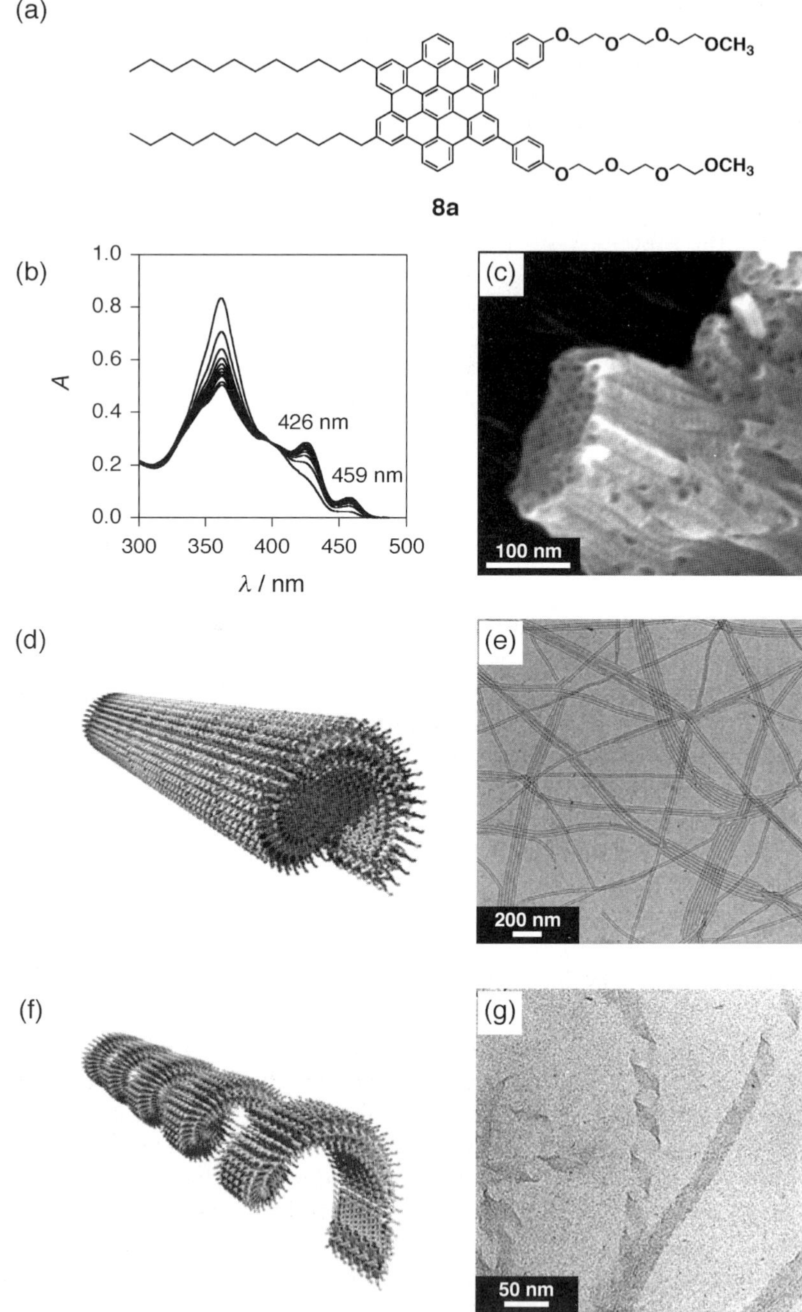
(a)
OCH_3
OCH_3
8a
(b)
1.0
0.8
0.6
A
0.4
0.2
0.0
426 nm
459 nm
300 350 400 450 500
λ / nm
(c)
100 nm
(d)
(e)
200 nm
(f)
(g)
50 nm

Fig. 8 Molecular structure of **8a** (**a**). Electronic absorption spectral change of **8a** upon self-assembly in THF (1 mg ml^{-1}) (**b**). SEM micrograph (**c**), proposed structure (**d**), and TEM micrograph (**e**) of tubularly assembled **8a**. Proposed structure of the nanocoils of self-assembled **8a** (**f**). TEM micrograph of a mixture of nanotubes and nanocoils formed by the self-assembly of **8a** in a mixture of THF and water (8/2 v/v) (1 mg ml^{-1})

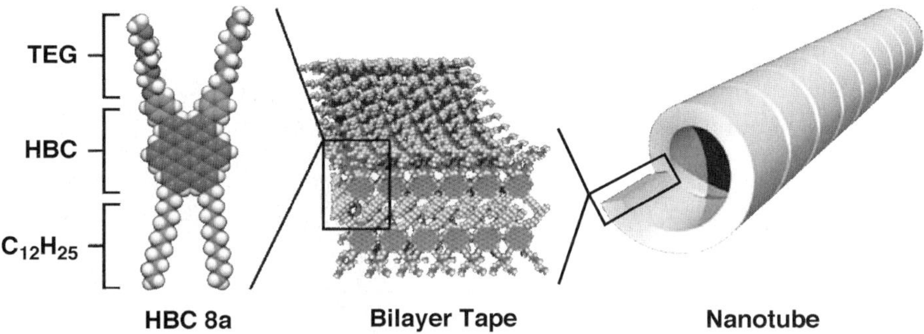

Fig. 9 Proposed molecular arrangement at the cross-section of the nanotube of **8a**

The nanotubular assembly of **8a** is a substantial insulator. However, since HBC derivatives are redox active [30], charge carriers can be generated in the nanotubes upon oxidation with, e.g., NOBF$_4$. A conductivity measurement using nano-gap (180 nm) electrodes allows detection of the conducting behavior of a single piece of the doped nanotube, indicating that a great number of the HBC units are electronically coupled in the "graphite wall" to provide a carrier-transport pathway. As evaluated by the slope of the observed linear I–V profile, the resistivity at 285 K is 2.5 MΩ, which increases as the temperature is lowered.

4.1.2
Covalently Stabilized HBC Nanotubes

Molecular self-assembly has been recognized as a powerful approach to designer soft materials with a nanoscopic structural precision [11]. However, self-assembled nanostructures are inherently subject to disruption with heating and exposure to solvents. The HBC nanotubes are not exceptional. Thus, for practical applications of the nanotubes, one has to consider post-modification of their nanostructures for covalent connection of the assembled HBC units. Because the inner and outer surfaces of the nanotubes are covered with TEG chains, incorporation of a polymerizable functionality into the TEG termini allows for the formation of surface polymerized nanotubes with an enhanced morphological stability.

4.1.2.1
Surface Polymerization via Olefin Metathesis

Olefin metathesis is a highly efficient carbon–carbon bond-forming reaction usable for the synthesis of polymeric materials. Since the discovery of Grubbs catalysts [31], a wide variety of polymers with controlled architectures have been prepared through acyclic diene metathesis (ADMET) [32] and ring-opening metathesis polymerizations (ROMP) [33]. In order to obtain covalently stabilized HBC nanotubes, allyl group-appended HBC **8b** (Fig. 10) is designed for post-ADMET [34]. Under conditions similar to those for the self-assembly of **8a**, HBC **8b** self-assembles into nanotubes. However, the attempted ADMET using the first-generation Grubbs catalyst occurs only sluggishly (Fig. 11). In contrast, ADMET proceeds when the catalyst is added to a homogeneous CH_2Cl_2 solution of **8b**. Unexpectedly, the reaction of unassembled **8b** affords nanotubes quantitatively (Fig. 11). Since the

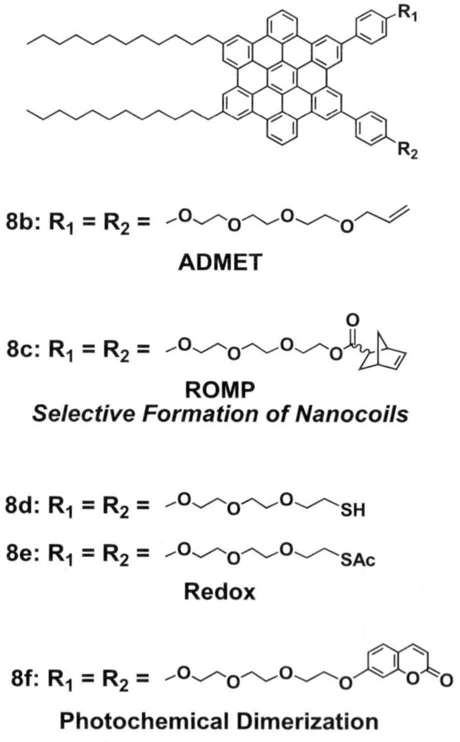
R1
R2
8b: R1 = R2 =
ADMET
8c: R1 = R2 =
ROMP
Selective Formation of Nanocoils
8d: R1 = R2 =
SH
8e: R1 = R2 =
SAc
Redox
8f: R1 = R2 =
Photochemical Dimerization

Fig. 10 Molecular structures of **8b-8f**

Fig. 10 Molecular structures of **8b–8f**

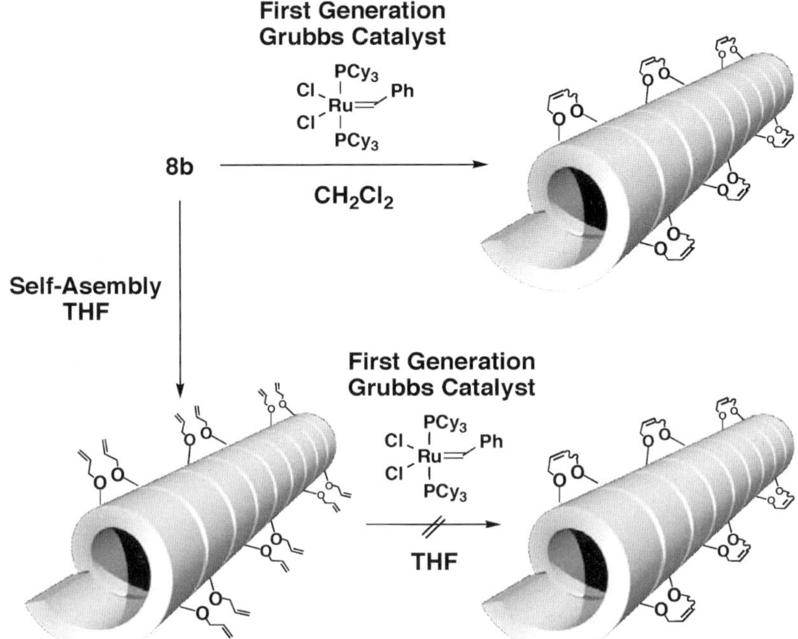

Fig. 11 Synthetic approaches to the surface-polymerized nanotube of **8b** by acyclic diene metathesis (ADMET) polymerization

nanotubes thus obtained are surface polymerized, they show an enhanced thermal stability. While the nonpolymerized nanotube displays a softening temperature of 195 °C, that for the polymerized one is 244 °C. Furthermore, upon heating at 175 °C, most of the polymerized nanotubes survive even after 24 h, whereas the non-polymerized ones are completely disrupted within 2 h. The polymerized nanotubes are highly insoluble and can preserve the hollow structure upon immersion in organic solvents.

Differing from **8a** and **8b**, norbornene-appended HBC **8c** (Fig. 10), designed for ROMP, affords nanocoils as well as nanotubes (Fig. 12) [35]. By choosing appropriate conditions, either nanocoils or nanotubes are selectively formed. For example, by Et_2O vapor diffusion into a CH_2Cl_2 solution of **8c** (0.65 mM) at 15 °C, the nanocoils form exclusively. At the same time, when the vapor diffusion is conducted at 25 °C at a lower concentration of **8c** (0.22 mM), the nanotubes can be obtained as the sole product. As observed by TEM and SEM, the nanocoil consists of a 20-nm-wide bilayer tape with 30 nm in diameter and 60 nm in pitch (Fig. 13a), while the nanotube possesses a diameter of 20 nm and a wall thickness of 3 nm (Fig. 13b). Interestingly, when the suspension of the nanocoils is allowed to stand for 5 days at 25 °C, the coiled structure quantitatively transforms into the tubular morph-

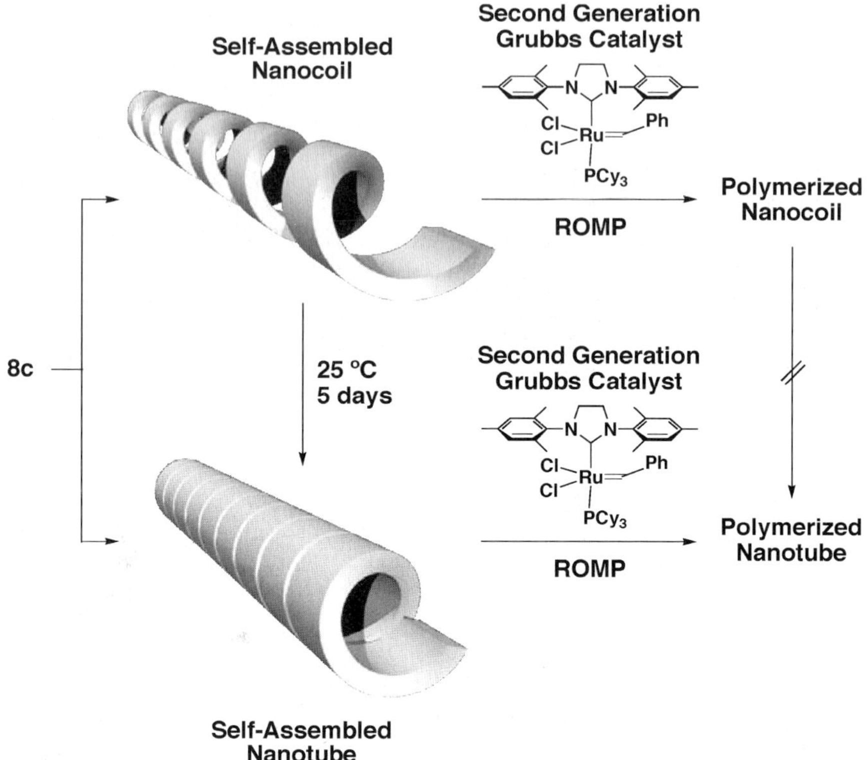

Fig. 12 Synthetic approaches to the surface-polymerized nanotube and nanocoil of **8c** by post-ring-opening metathesis polymerization (ROMP)

ology (Fig. 13c), indicating that the nanocoil is the kinetic intermediate for the nanotube.

Post-ROMP of the norbornene functionality enables stabilization of the thermodynamically metastable nanocoil. Thus, when the second generation Grubbs catalyst is added at $20\,^\circ\text{C}$ to an Et_2O/CH_2Cl_2 (100/1 v/v) suspension of the nanocoils of **8c**, a polymeric substance forms. Infrared spectroscopy indicates the transformation of the cyclo-olefinic C=C bond of **8c** into an acyclic one in 80% yield. TEM and SEM micrographs reveal that the nanocoils after ROMP preserve the helical structure and size regime (Fig. 13d). Similar to polymerized nanotubes of **8b**, the nanocoils after ROMP are thermally stable. For instance, the coiled structure survives at $75\,^\circ\text{C}$ even after 12 h in Et_2O/CH_2Cl_2 (100/1 v/v), while the non-polymerized coils are completely disrupted. The polymerized nanocoils are insoluble in good solvents for monomeric **8c**. Likewise, post-ROMP occurs on the nanotubes of assembled **8c** and enhances their physical robustness.

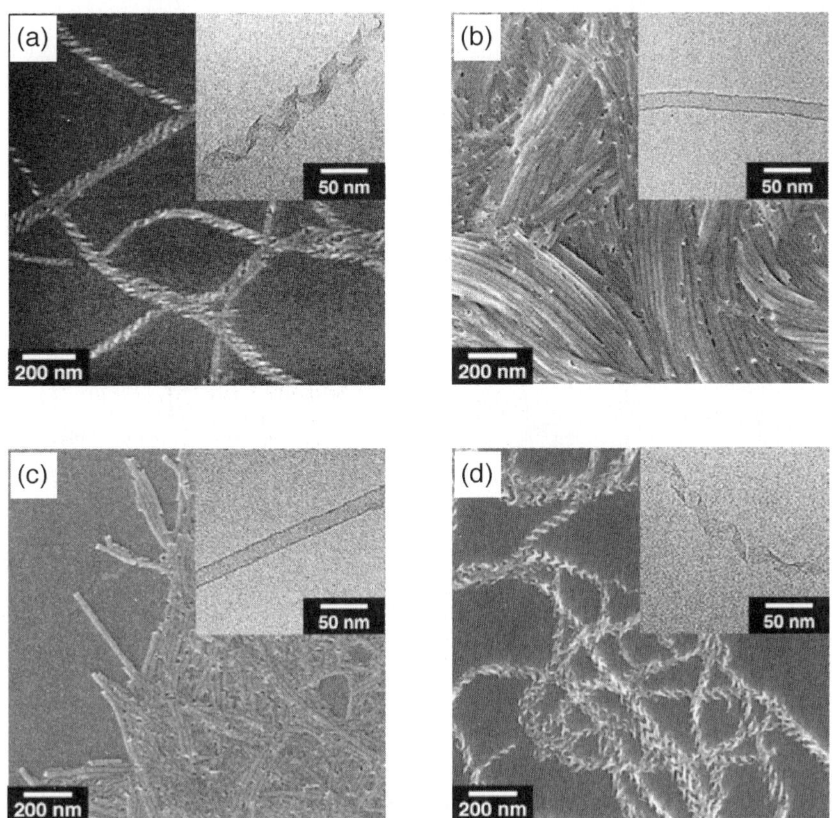

Fig. 13 SEM and TEM (*inset*) micrographs of the nanocoils (**a**) and nanotubes (**b**) of self-assembled **8c**, formed by Et$_2$O vapor diffusion into CH$_2$Cl$_2$ solutions of **8c**. SEM and TEM (*inset*) micrographs of the nanotubes of self-assembled **8c**, formed upon immersion of the nanocoils in Et$_2$O/CH$_2$Cl$_2$ (9/1) at 25 °C for 5 days (**c**). SEM and TEM (*inset*) micrographs of the surface-polymerized nanocoils of **8c** formed by post-ROMP (**d**)

4.1.2.2
Oxidative and Photochemical Surface Polymerizations

HBC nanotubes that undergo a reversible surface polymerization by external stimuli are interesting, as they can be used for lithographic patterning. Along this line, HBCs **8d** [36] and **8f** [37], appended with thiol and coumarin groups, respectively, are designed (Fig. 10), which can be stitched in a self-assembled state by post-polymerization via redox and photochemical intermolecular dimerization, respectively. For example, acetyl-protected thiol-appended HBC **8e** (Fig. 10) self-assembles in THF to form nanotubes with a diameter of 20 nm and a wall thickness of 3 nm (Fig. 14). Deprotection of the acetyl groups of tubularly assembled **8e** with NaOEt under aerobic conditions affords a solid

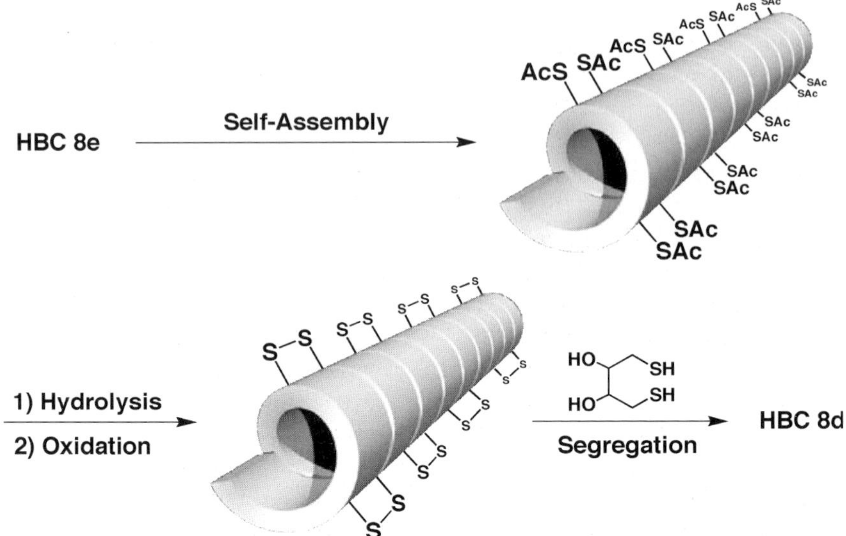

Fig. 14 Schematic representations of the oxidative polymerization and reductive depolymerization on the nanotube of self-assembled **8e**

(a)

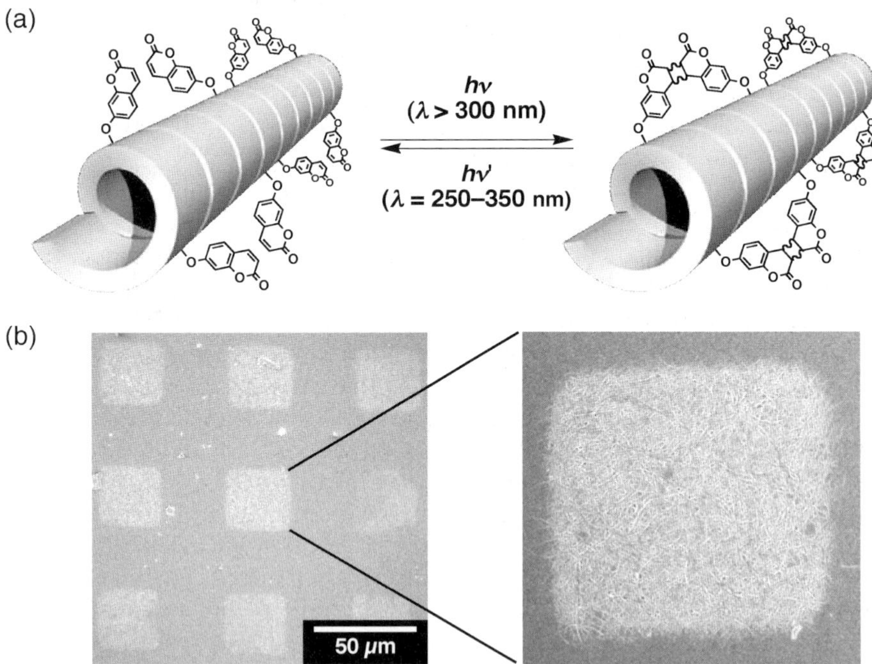

(b)

Fig. 15 Schematic representations of the photochemical polymerization and depolymerization on the nanotube of self-assembled **8f** (**a**). SEM micrographs of a negative pattern developed by photochemical polymerization followed by rinsing (**b**)

substance, which is insoluble in common organic solvents such as CH_2Cl_2, $CHCl_3$, and THF. Although partial defects are detected, the tubular morphology is preserved after the deprotection, suggesting the formation of disulfide bonds on the nanotubes. When the disulfide bonds are reduced with dithiothreitol in refluxing THF, monomeric **8d** is recovered.

Coumarin derivatives are known to dimerize reversibly by choosing suitable irradiation wavelengths [38]. Thus, utilization of coumarin allows photochemical stitching and unstitching of tubularly assembled HBC under dry conditions (Fig. 15a). HBC **8f** (Fig. 10), designed for this purpose, self-assembles into nanotubes under selected conditions. For example, when EtOH vapor is allowed to diffuse at 25 °C into a $CHCl_3$ solution of **8f**, a nanotubular assembly forms, whose diameter and wall thickness are similar to those of the nanotube from **8a**. Upon exposure at 25 °C to a light with $\lambda > 300$ nm for 10 min, the nanotubes undergo photochemical dimerization of the coumarin pendants. Matrix-assisted laser desorption ionization time-of-flight (MALDI-TOF) mass spectrometry of the irradiated sample allows for the detection of shortchain oligomers of **8f**. By means of infrared spectroscopy, the conversion of the coumarin units is estimated as 20%. The irradiated sample is insoluble in $CHCl_3$, a good solvent for **8f**. As confirmed by SEM and TEM microscopy, the polymerized nanotubes preserve their hollow structure even after immersion in $CHCl_3$. Meanwhile, photochemical unstitching occurs readily by exposing an EtOH suspension of the surface-polymerized nanotubes to a shorter-wavelength UV light ($\lambda = 250-350$ nm). Upon dilution of the resulting suspension with $CHCl_3$, the nanotubes completely disassemble, affording a homogeneous solution of 8f.

By taking advantage of this reversible solubility change, both negative and positive patterns of the nanotubes can be developed by a lithographic post processing. For example, a metal grid is place on a cast film of the nanotubes for masking, and a light with $\lambda > 300$ nm is used for stitching the nanotubes located at the unmasked areas. Rinsing the resulting film with $CHCl_3$ allows selective removal of unpolymerized nanotubes, leaving a negative pattern (Fig. 15b). For positive patterning, the entire cast film is first exposed to a light with $\lambda > 300$ nm, and the resulting film, comprised of the stitched nanotubes, is covered with a metal grid. By subsequent irradiation of the sample using a light with $\lambda = 250-350$ nm and rinsing with $CHCl_3$, a positive pattern is developed on the substrate.

4.1.3
Stereochemical Aspects

All the HBC nanotubes and nanocoils described above are composed of helically rolled-up bilayer tapes and possess a supramolecular helical chirality as an essential structural element. This is quite interesting, since most HBC molecules examined are devoid of stereogenic centers. Conductive nanotubes and nanocoils with one-handed helical chirality have the potential for exploring electromagnetic properties. The following section features design and assembly of chiral HBC derivatives.

4.1.3.1
HBC Nanotubes with One-Handed Helical Chirality

For the control of the helical chirality of the nanotubes, chiral HBCs (*S*)-**8g** and (*R*)-**8g** (Fig. 16), carrying asymmetric centers in the hydrophilic TEG chains, are designed (Fig. 17) [39]. Controlled self-assembly of **8g** proceeds in 2-methyltetrahydrofuran (2-MeTHF) to give nanotubes with a size regime identical to that of the nanotube from achiral **8a**. For instance, a 2-MeTHF solution of (*S*)-**8g** (3 mg ml^{-1}), upon heating followed by cooling, shows a time-dependent spectral change profile, displaying red-shifted absorption bands at 398 and 421 nm, characteristic of tubularly assembled HBCs. The resulting suspension containing the nanotubes shows positive circular dichroism (CD) bands at 389, 400, and 423 nm (Fig. 18a), while unassembled (*S*)-**8g** in hot 2-MeTHF is CDsilent. The CD spectral change of (*R*)-**8g** upon self-assembly is a mirror image of that observed for (*S*)-**8g** (Fig. 18b), indicating the formation of nanotubes with an opposite handedness. Thus, the point chirality of **8g** is successfully translated into the supramolecular helical chirality of the nanotubes. In relation to these observations, self-assembly of **8g** by diffusion of hexane vapor into its chlorocyclohexane solution leads to the formation of helical nanocoils along with nanotubes. TEM analysis of the helical coils

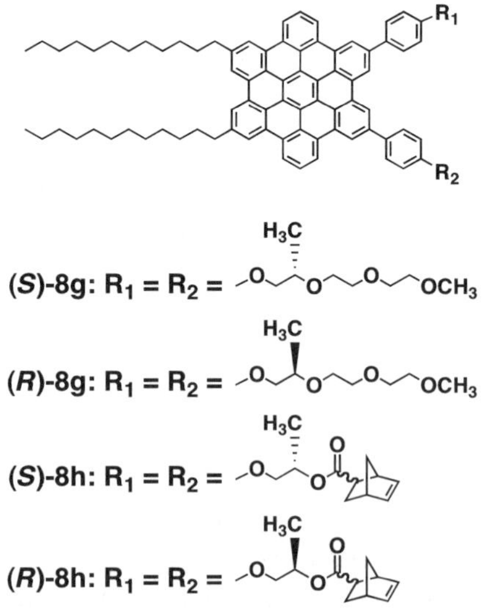

Fig. 16 Molecular structures of **8g** and **8h**

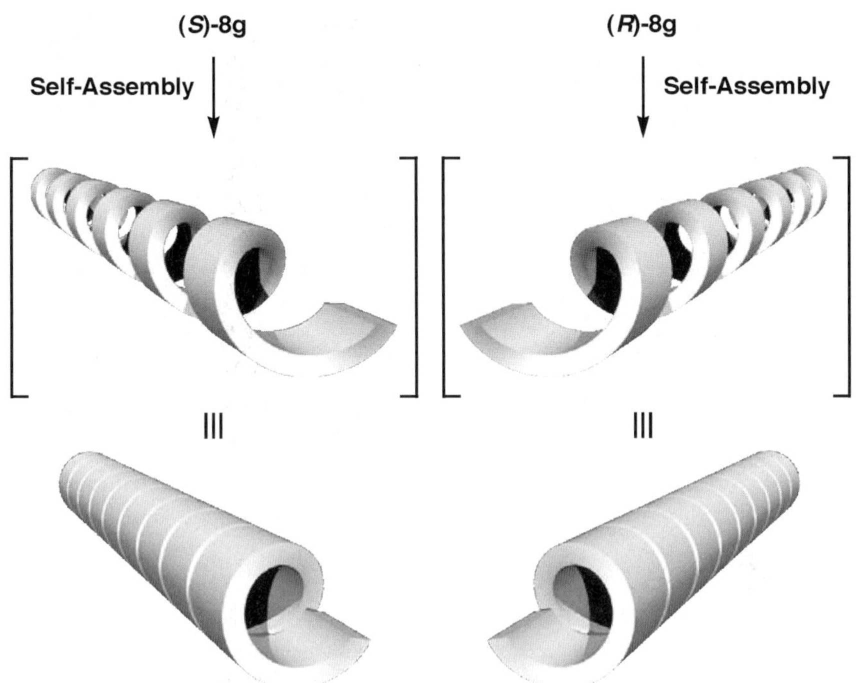

Fig. 17 Schematic representations of the formation of the one-handed nanotubes from chiral **8g**

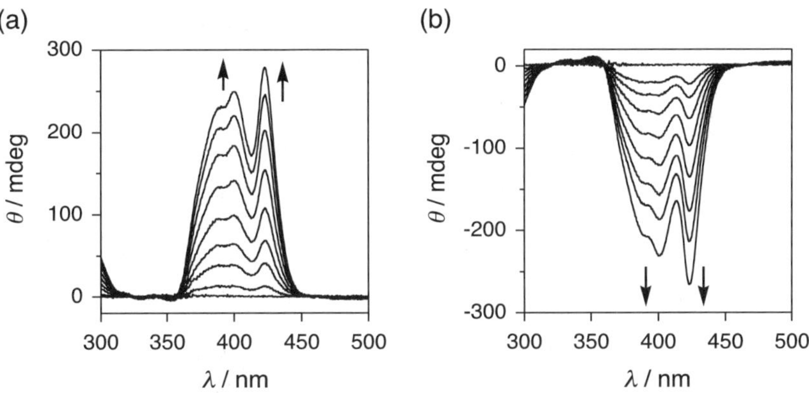

Fig. 18 Time-dependent CD spectral changes of (S)-**8g** (**a**) and (R)-**8g** (**b**) upon self-assembly in MeTHF (3 mg ml^{-1})

demonstrates nearly a 100% helical dominance (Fig. 19). Since the nanocoils are a precursor for the nanotubes, the above observation demonstrates that the nanotubes from **8g** are one-handed.

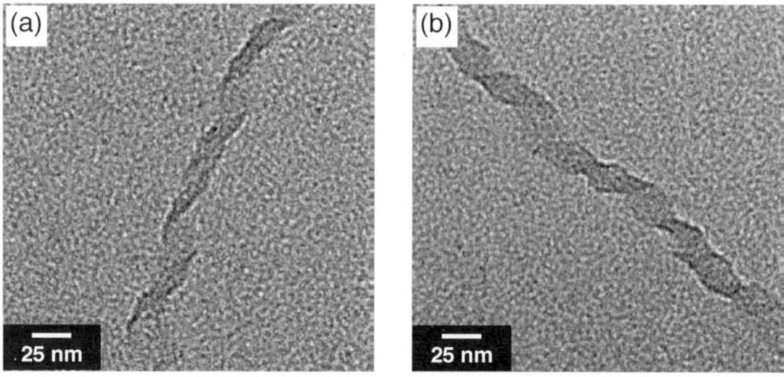

Fig. 19 TEM micrographs of the nanocoils of self-assembled (*S*)-**8g** (**a**) and (*R*)-**8g** (**b**)

Chiral amplification via the majority rule is a known phenomenon for covalent helical polymers [16, 17, 40, 41] but has not been reported for supramolecular assemblies until quite recently. In 2005, the group of Meijer has reported the first example of this in a supramolecular assembly [42]. In the same year, the authors' group has reported that the stereochemical aspect of the co-assembly of the (*S*)- and (*R*)-enantiomers of HBC **8g** follows the majority rule [39]. Controlled assembly of **8g** takes place in 2-MeTHF at varying mole ratios of the (*S*)- and (*R*)-enantiomers, affording high-quality nanotubes. Plots of the CD intensity at 423 nm of the resulting nanotubes versus the enantiomeric excess (ee) of **8g** show a sigmoidal feature (Fig. 20), indicating that the two enantiomers indeed co-assemble, where the helical handedness of the nanotubes is determined by the major enantiomer.

The nanotubes from **8g** are much longer than those from other HBC derivatives, and can be aligned unidirectionally [43]. When (*S*)-**8g** in

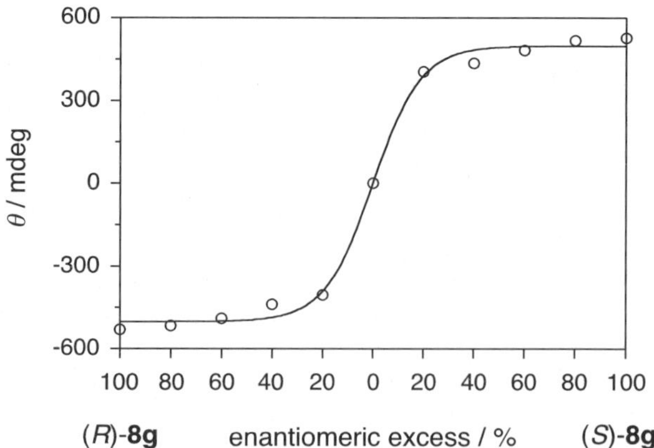

Fig. 20 Plots of the CD intensity at 423 nm versus the enantiomeric excess of **8g**

2-MeTHF is allowed to stand for three weeks at 25 °C, a suspension containing bundles of several hundred-micrometer-long nanotubes results. When a glass hook is dipped into this suspension repeatedly, the bundled nanotubes are collected, which are then pulled up to form a macroscopic fiber. By way of this simple treatment, most of the nanotubes are aligned unidirectionally along the longer axis of the macroscopic fiber. When this macroscopic fiber is doped with I_2, the resistivity decreases from 200 MΩ cm to 20 Ω cm at 300 K. Since the resistivity increases upon lowering the temperature, the doped fiber is a semiconductor. Due to the unidirectional alignment of the nanotubes, the fiber shows an anisotropic electrical conduction, where the resistivity along the fiber axis at 55 K is 1/35 as small as that along its perpendicular direction.

4.1.3.2
HBC Nanocoils with One-Handed Helical Chirality

As described in the above section, the self-assembly of chiral **8g** affords nanotubes with one-handed helical chirality. However, in regard to the exploration of electromagnetic properties, one-handed coils are more interesting. Section 4.1.2.1 highlights the formation of nanocoils from HBC **8c**, which is stereochemically non-selective, affording a mixture of right-and lefthanded nanocoils. Recently, the authors have reported that **8c** in conjunction with **8h** (Fig. 16), under selected conditions, assembles into nanocoils with one-handed helical chirality [44]. When Et_2O vapor is allowed to diffuse at 15 °C into a CH_2Cl_2 solution of a mixture of **8c** (0.52 µmol, 80 mol %) and (*S*)-**8h** (0.13 µmol, 20 mol %), a yellow precipitate forms quantitatively after 24 h. SEM microscopy of the precipitate displays only left-handed nanocoils with a diameter of 30 nm and a pitch of 60 nm. By means of post-ROMP of the norbornene pendants, this morphology is covalently fixed without any

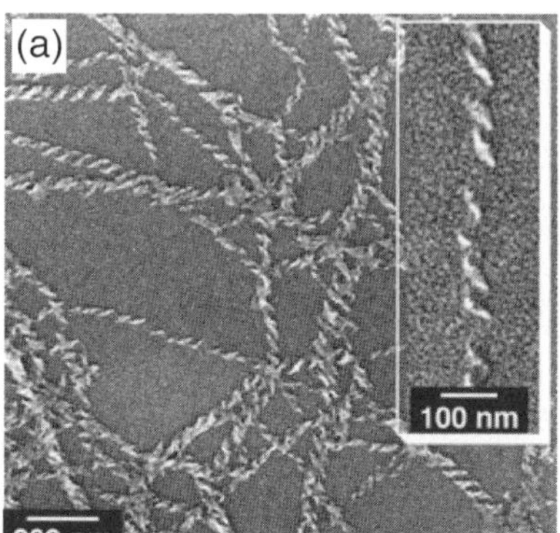

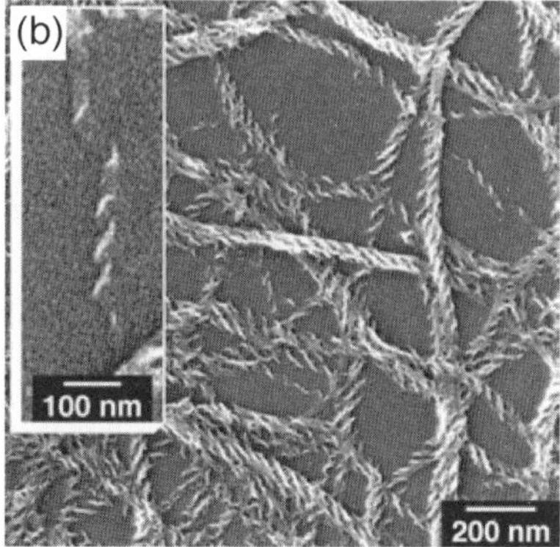

Fig. 21 SEM micrographs of the surface-polymerized one-handed nanocoils of **8c** with 20 mol % of (*S*)-**8h** (**a**) and 20 mol % of (*R*)-**8h** (**b**)

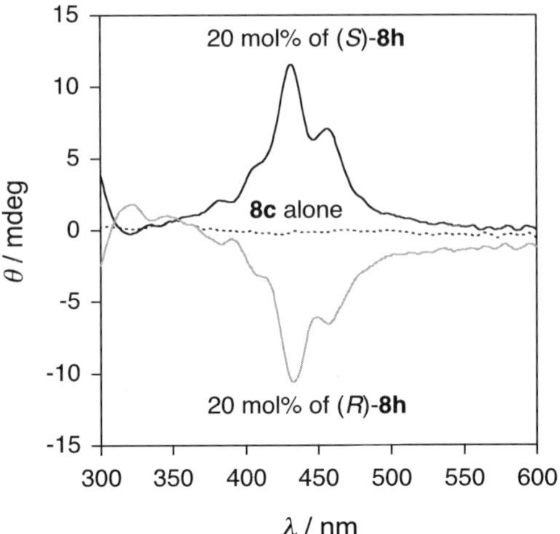

Fig. 22 CD spectra of the surface-polymerized nanocoils of **8c** with 20 mol % of (S)-**8h**, and 20 mol % of (R)-**8h**

disruption (Fig. 21a). As expected, the use of (R)-**8h** instead of (S)-**8h** for the co-assembly with **8c** followed by ROMP exclusively affords right-handed nanocoils (Fig. 21b), which display a mirror-image CD spectrum of the left-handed nanocoils (Fig. 22). The nanocoils are one-handed when the mole fraction of **8h** is in a range from 20 to 50 mol %. As described in Sect. 2.2, these observations can be accounted for by the sergeants-and-soldiers effect. However, an interesting aspect is that the soldier (**8c**) in this co-assembling system has a preference to form a coiled architecture, but the sergeant (**8h**), which is responsible for determining the handedness of the coil, just gives microfibers upon self-assembly. Nevertheless, to ensure the selective formation of one-handed nanocoils, a very delicate optimization of the composition is required. For example, when the mole fraction of **8h** exceeds 50 mol %, nanotubes begin to form concomitantly with nanocoils, while neither **8c** nor **8h** assembles into nanotubes under the conditions examined.

4.2
Charged Polycyclic Aromatic Hydrocarbons

Müllen et al. have reported an interesting observation that polycyclic aromatic hydrocarbon **9** (Fig. 23a) containing a positively charged nitrogen atom yields either nanotubes or ribbon-like aggregates, depending on the counteranion [45]. When a MeOH solution of **9a** is drop cast, ribbon-like structures with a width of 80 nm form, as observed by TEM and SEM microscopy. Wide-angle X-ray scattering (WAXS) confirms that the ribbons are comprised of a lamellar structure. It is of interest that when the counteranion of **9a** is changed from Cl^- to BF_4^-, a drop-cast film from MeOH contains both

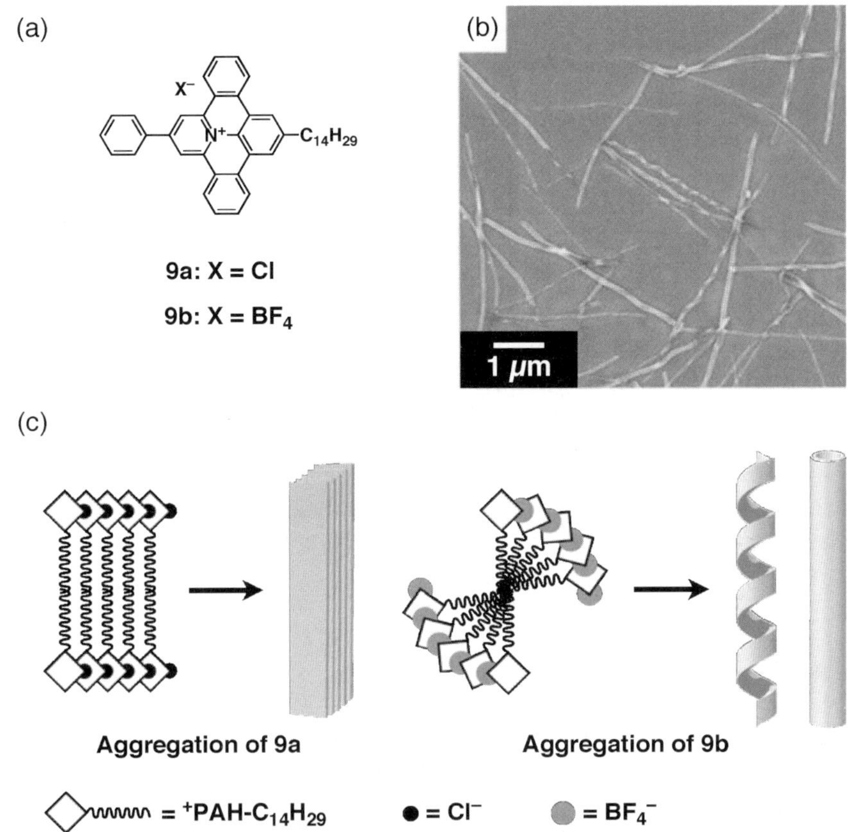

Fig. 23 Molecular structures of **9a** and **9b** (**a**). SEM micrograph of a mixture of the nanotubes and nanocoils of self-assembled **9b**. Reproduced with permission from [45] (**b**). Proposed mechanism of the formation of the ribbon from **9a** and the nanotube from **9b** (**c**)

nanocoils and nanotubes. SEM microscopy shows that the nanocoils possess varying pitches while the nanotubes are, on average, 5-μm-long with a diameter of 80–150 nm (Fig. 23b). TEM analysis reveals that the inner diameter and wall thickness of the nanotubes are 20–50 and 40–60 nm, respectively. According to WAXS profiles, both the ribbons from **9a** and the nanotubes from **9b** consist of a lamellar structure, but the packing geometries are different from one another. The authors suggest that the steric effect of the counteranion of **9** plays a major role in selecting the morphology. In the ribbon of **9a**, the positively charged aromatic parts are oriented cofacially to one another and sandwich a Cl⁻ ion between them (Fig. 23c). At the same time, in the nanocoil or nanotube of **9b**, the aromatic parts are considered to adopt an offset geometry to one another. The aromatic parts may hardly orient cofacially probably due to the bulky BF_4^- ion.

5
Perspectives

Until recently, only biologically relevant molecules such as lipids and peptides were known to form nanotubes and nanocoils [46]. However, as described in this review article, successful examples have gradually been extended to π-conjugated molecules. Because of an increasing attention to organic electronics, the research field on electroactive nanotubes has begun to emerge. In principle, tubes are quasi-equivalent to two-dimensional thin films. However, it is possible that they allow directional transports of energy and charge carriers. Tubes are able to accommodate functional guest molecules that can collaborate to provide synergistic functions. Nevertheless, despite such a greater fascination with them over more abundant fibers and rods, successful examples are still fewer, resulting in only a limited knowledge about their rational molecular design [47]. Through this review article, one may notice that chirality might be a potent tool, as it possibly gives rise to a twisted (helical) geometry of the assembled π-conjugated components, which is favorable for two-dimensional objects to roll up or coil.

References

1. Iijima S (1991) Nature 354:56
2. Iijima S, Ichihashi T (1993) Nature 363:603
3. Jorio A, Dresselhaus G, Dresselhaus MS (eds) (2008) Carbon Nanotubes: Advanced topics in the synthesis, structure, properties and applications. Springer, New York
4. Yanagi K, Miyata Y, Kataura H (2008) Appl Phys Express 1:034003
5. Oda R, Huc I, Schmutz M, Candau SJ, MacKintosh FC (1999) Nature 399:566
6. Brizard A, Aimé C, Labrot T, Huc I, Berthier D, Artzner F, Desbat B, Oda R (2007) J Am Chem Soc 129:3754
7. Würthner F, Chen Z, Hoeben FJM, Osswald P, You C-C, Jonkheijm P, Van Herrikhuyzen J, Schenning APHJ, Van der Schoot PPAM, Meijer EW, Beckers EHA, Meskers SCJ, Janssen RAJ (2004) J Am Chem Soc 126:10611
8. Messmore BW, Sukerkar PA, Stupp SI (2005) J Am Chem Soc 127:7992
9. Bae J, Choi J-H, Yoo Y-S, Oh N-K, Kim B-S, Lee M (2005) J Am Chem Soc 127:9668
10. Schenning APHJ, Meijer EW (2005) Chem Commun 3245
11. Hoeben FJM, Jonkheijm P, Meijer EW, Schenning APHJ (2005) Chem Rev 105:1491
12. Yang W-Y, Lee E, Lee M (2006) J Am Chem Soc 128:3484
13. Ryu J-H, Oh N-K, Lee M (2005) Chem Commun 1770
14. Ajayaghosh A, Varghese R, George SJ, Vijayakumar C (2006) Angew Chem Int Ed 45:1141
15. George SJ, Ajayaghosh A (2005) Chem Eur J 11:3217
16. Green MM, Park J-W, Sato T, Teramoto A, Lifson S, Selinger RLB, Selinger JV (1999) Angew Chem Int Ed 38:3138
17. Yashima E, Maeda K, Nishimura T (2004) Chem Eur J 10:42
18. Ajayaghosh A, Varghese R, Mahesh S, Praveen VK (2006) Angew Chem Int Ed 45:7729
19. Ajayaghosh A, Varghese R, Praveen VK, Mahesh S (2006) Angew Chem Int Ed 45:3261

20. Elemans JAAW, Van Hameren R, Nolte RJM, Rowan AE (2006) Adv Mater 18:1251
21. Wang Z, Medforth CJ, Shelnutt JA (2004) J Am Chem Soc 126:15954
22. Sly J, Kasák P, Gomar-Nadal E, Rovira C, Górriz L, Thordarson P, Amabilino DB, Rowan AE, Nolte RJM, (2005) Chem Commun 1255
23. Watson MD, Fechtenkötter F, Müllen K (2001) Chem Rev 101:1267
24. Clar E, Ironside CT, Zander M (1959) J Chem Soc 142
25. Van de Craats AM, Warman JM, Müllen K, Geerts Y, Brand JD (1998) Adv Mater 10:36
26. Schmidt-Mende L, Fechtenkötter A, Müllen K, Moons E, Friend RH, MacKenzie JD (2001) Science 293:1119
27. Simpson CD, Wu J, Watson MD, Müllen K (2004) J Mater Chem 14:494
28. Wu J, Pisula W, Müllen K (2007) Chem Rev 107:718
29. Hill JP, Jin W, Kosaka A, Fukushima T, Ichihara H, Shimomura T, Ito K, Hashizume T, Ishii N, Aida T (2004) Science 304:1481
30. Rathore R, Burns CL (2003) J Org Chem 68:4071
31. Trnka TM, Grubbs RH (2001) Acc Chem Res 34:18
32. Davidson T, Wagener KB (1999) Acyclic Diene Metathesis (ADMET) Polymerization. In: Schluter AD (ed) Synthesis of Polymers. Materials Science and Technology: Synthesis of Polymers, Chap 4. Wiley-VCH, New York
33. Grubbs RH (2006) Angew Chem Int Ed 45:3760
34. Jin W, Fukushima T, Kosaka A, Niki M, Ishii N, Aida T (2005) J Am Chem Soc 127:8284
35. Yamamoto T, Fukushima T, Yamamoto Y, Kosaka A, Jin W, Ishii N, Aida T (2006) J Am Chem Soc 128:14337
36. Motoyanagi J, Fukushima T, Kosaka A, Ishii N, Aida T (2006) J Polym Sci A 44:5120
37. Motoyanagi J, Fukushima T, Ishii N, Aida T (2006) J Am Chem Soc 128:4220
38. Trenor SR, Shultz AR, Love BJ, Long TE (2004) Chem Rev 104:3059
39. Jin W, Fukushima T, Niki M, Kosaka A, Ishii N, Aida T (2005) Proc Natl Acad Sci USA 102:10801
40. Nakano T, Okamoto Y (2001) Chem Rev 101:4013
41. Van Gestel J (2004) Macromolecules 37:3894
42. Van Gestel J, Palmans ARA, Titulaer B, Vekemans JAJM, Meijer EW (2005) J Am Chem Soc 127:5490
43. Yamamoto Y, Fukushima T, Jin W, Kosaka A, Hara T, Nakamura T, Saeki A, Seki S, Tagawa S, Aida T (2006) Adv Mater 18:1297
44. Yamamoto T, Fukushima T, Kosaka A, Jin W, Yamamoto Y, Ishii N, Aida T (2008) Angew Chem Int Ed 47:1872
45. Wu D, Zhi L, Bodwell GJ, Cui G, Tsao N, Müllen K (2007) Angew Chem Int Ed 46:5417
46. Shimizu T, Masuda M, Minamikawa H (2005) Chem Rev 105:1401
47. Shimizu T (2008) J Polym Sci A 46:2601

Adv Polym Sci (2008) 220: 29–64
DOI 10.1007/12_2008_159
© Springer-Verlag Berlin Heidelberg
Published online: 5 July 2008

Block Copolymer Nanotubes Derived from Self-Assembly

Guojun Liu

Department of Chemistry, Queen's University, 90 Bader Lane,
Kingston, Ontario K7L 3N6, Canada
guojun.liu@chem.queensu.ca

Abstract Block copolymer nanotubes are discrete cylindrical structures with a tubular core made from block copolymers. The diameter of such cores should be below 100 nm, and the length of the tubes can be up to hundreds of micrometers or longer. This article discusses their preparation, dilute solution properties, and chemical reactions.

1
Introduction

Block copolymer nanotubes in this chapter refer to discrete cylindrical structures with a tubular core diameter ranging from a few nanometers to 100 nm and a length-to-diameter aspect ratio substantially larger than 1. The focus of this review is on discrete nanotubes that can be dispersed as colloidal entities in a solvent. Thus, parallel arrays of nanotubes that have been prepared by chemical processing is not the topics of this article. Examples of these types of systems include those involving the selective cylindrical domain degradation and possibly crosslinking of the matrix phase of block-segregated copolymer thin films. Such thin films have been used as membranes [1–5], as nanolithographic masks [5–9], and as the templates for metal nanorod preparation [10]. Discrete nanotubes have also been prepared from sacrificial templates. In approach 1, homopolymer or block copolymer nanotubes are prepared using the tubular pores of anodized alumina or ion-track-etched membranes as the template [11]. Such preparation involved first sucking a polymer solution into the template pores, then evaporating the solvent to collapse the polymer on the wall of the template, and finally etching away the template leaving the individual tubes behind [12]. In approach 2, discrete cylindrical or tubular templates are used. This involves first the deposition of polymer on the template surface relying on specific interactions such as electrostatic attraction and H-bonding, and then the etching of the template to yield nanotubes [13]. The preparation and study of templated nanotubes are reviewed in another article in this volume (Steinhart 2008, this volume). The focus of discussion here is on discrete or solvent-dispersible nanotubes derived from the self-assembly of block copolymers. For a comprehensive review of nanotubes derived from surfactants and chiral phospholipids, etc., readers are referred to an excellent review by Shimizu et al. [14].

Figure 1 depicts structures of nanotubes that have so far been derived from block copolymer self-assembly. While the nanotubes are drawn as being rigid and straight, they, in reality, can bend or contain kinks. The top scheme depicts a nanotube formed from either an AB diblock copolymer [15, 16] or an ABA triblock copolymer [17], where the gray B block forms a dense intermediate shell and the dark A block or A blocks stretch into the solvent phase from both the inner and outer surfaces of the gray tubular shell. Such tubes have been prepared so far from the direct self-assembly or tubular micelle formation of a few block copolymers in block-selective solvents, which solubilize only the dark A block or blocks. Nanotubes with structures depicted in the middle and bottom schemes have been prepared from precursory ABC triblock copolymer nanofibers, which consist of an A corona, a cross-linked intermediate B shell, and a C core [18] A fully empty tubular core was ob-

Fig. 1 Structures of block copolymer nanotubes prepared so far

tained by degrading completely the innermost C block to yield nanotubes with structure depicted in the middle scheme of Fig. 1 [19]. Tubular cores lined by a third polymer as depicted in the bottom scheme of Fig. 1 were obtained from cleaving pendant groups off the core block of an ABC triblock copolymer nanofiber [20, 21].

Overall, reports on preparation of nanotubes from block copolymers have been rare, and there have been no reports on practical applications of such structures. For this, the emphasis of this chapter will be on the fundamental aspects of these materials. In Sect. 2, nanotube or tubular micelle formation from the direct self-assembly of block copolymers in block-selective solvents will be reviewed. Section 3 will be mainly on nanotubes derived from the chemical processing of cross-linked triblock copolymer nanofibers. Example nanotube preparations will be given, dilute solution properties of the nanotubes will be discussed, and the different reaction patterns of the nanotubes will be examined. Concluding remarks will be made in Sect. 4.

2
Self-Assembled Nanotubes

2.1
Block Copolymer Self-Assembly in Block-Selective Solvents

In a block-selective solvent, the insoluble block or blocks of a copolymer agglomerate to form nanometer-sized aggregates [5, 22, 23]. Such aggregates disperse in the solvent and are protected from further agglomeration by the soluble block(s) that form(s) the aggregate corona.

The aggregates formed can be micelles, which are structures with the lowest Gibbs free energy under a given set of experimental conditions. Depending on the temperature and the polymer molecular weight(s), the chain mobility of the aggregated or core block(s) can be limited. This may prevent the attainment of the micellar structures. Under such circumstances, the aggregates are formed as a meta-stable or kinetically trapped species. When one does not know the exact nature of an aggregate produced, it is normally referred to as a micelle-like aggregate or an aggregate.

Depending on the solvent, block copolymer composition, and preparation method, the micelle-like aggregates can have various morphologies [24–30]. For diblock aggregates, the morphologies range from spheres [23] to cylinders [31–34], vesicles or bilayers [35, 36] etc. Micelles with different morphologies form because they are the energetically most-favored products under their formation conditions. The free energy of a coil–coil diblock micelle has contributions from the stretching energies of the coronal and core chains and the energy for the interface between the core and solvent [22]. The quality of a block-selective solvent affects the morphology of the micelles because it changes the core/solvent interfacial energy. For experiments carried out in a series of solvents of decreasing quality for the core block, the core/solvent interfacial tension increases as the solvent for the core polymer deteriorates. This change in solvent quality is generally accompanied by an increase in micelle size and aggregation number, and accompanied by stretching of the core and coronal chains. Above a critical interfacial tension, a morphological transition occurs, e.g., a spherical-to-cylindrical micelle transition. Immediately after this transition, the size of the micelles decreases to reduce the stretching energies of the core and coronal chains [25, 37]. For the sphere-to-cylinder transition, it is the radius that contracts. For cylinder-to-platelette transition, the thickness of the platelette core becomes thinner than the radius of the precursory cylinder.

Diblock copolymers with different relative coronal and core chain lengths form micelles of different shapes again to minimize the system's free energy. Coil–coil diblocks with a long soluble block and short insoluble block tend to form spherical micelles because spherical micelles possess the largest core/coronal interfacial curvature. A large interfacial curvature helps to reduce the repulsion between the coronal chains and thus their stretching energy. Figure 2 compares the coronal and core chain packing in a cylindrical and bilayer micelle, respectively, and evidently a longer coronal chain can be packed into the coronal of the cylindrical micelle. Thus, micelles of a homologous series of coil–coil diblock copolymers normally undergo the morphological transition from spheres to cylinders and to bilayers (vesicles are enclosed bilayers), as the length of the soluble block is decreased relative to that of the insoluble block [25, 26].

Aside from block copolymer composition and the quality of the block-selective solvent, polymer concentration can also affect micellar morphology. For example, the length of cylindrical micelles normally increases with the concentration of diblock copolymers in a block-selective solvent [38–40]. If the aggregates are kinetic products, the morphologies will be affected by sample preparation conditions and history as well.

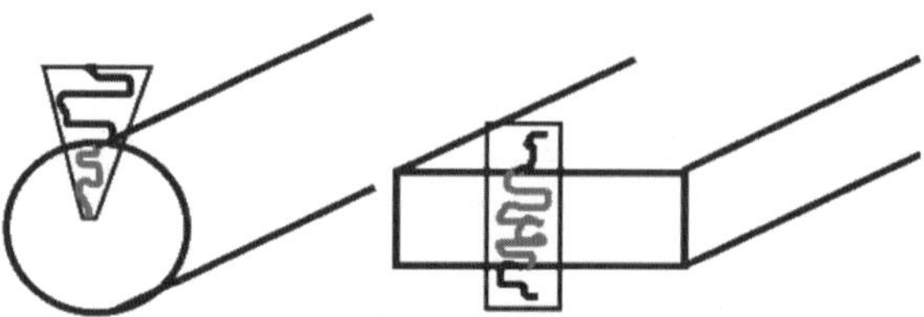

Fig. 2 Comparison of chain packing in diblock cylindrical and vesicular micelles

2.2
Self-Assembled Block Copolymer Nanotubes

Despite many reports on nanotube formation from phospholipids, glycolipids, peptidic amphiphiles, and other small-molecule surfactants and theoretical studies of this subject [14], reports on nanotube formation from the self-assembly of block copolymers are rare. There have been no theoretical treatments examining their formation or properties. It is not even known if block copolymer nanotubes are thermodynamically stable entities or kinetically controlled association products when formed in block-selective solvents.

The first example of block copolymer nanotube formation was reported by Eisenberg and coworkers [15, 16], who studied three polystyrene-*block*-poly(ethylene oxide) or PS-PEO samples with the styrene repeat unit number fixed at 240 and EO repeat units number varying between 15, 45, and 80, respectively. The micelle-like aggregates were prepared by the following protocol: First, the diblock copolymers were either dissolved molecularly in N,N-dimethylforamide (DMF) or dispersed in a mixture of DMF/water containing < 6.5 wt. % of water, a block-selective solvent for PEO. Then, water was added drop-wise to 25 wt. % to freeze in the aggregated structure. In the final step, DMF was removed from the solution by dialysis against water. Their systematic study revealed that nanotubes were formed readily under many preparation conditions from PS_{240}-PEO_{15}, where the subscripts 240 and 15 denote the repeat unit numbers for styrene and ethylene oxide, respectively. The optimal conditions to prepare nanotubes involved dispersing the diblock at a concentration of 2.0 wt. % in DMF/water containing 4.5 wt. % of water and equilibrating the mixture for a long period of time, e.g., 10 weeks, before performing the water addition and dialysis steps.

Figure 3A and B show transmission electron microscopy (TEM) images of nanoaggregates prepared from PS_{240}-PEO_{15} by the above protocol using the sample annealing times of 10 and 3 weeks in DMF/water containing 4.5 wt. % of water. The sample shown in Fig. 3B consisted of a mixture of tubes and vesicles, and only tubes are seen in Fig. 3A. This result suggests that the population of the nanotubes increased at the cost of vesicles (Fig. 3) as the sample annealing time in the initial DMF/water mixture increased. The tubular nature of the elongated structures was better seen from TEM images with higher magnifications. They showed clearly the higher transmission in the center part of the aggregates. The triblock chains should be packed in the fashion as depicted in the top Scheme of Fig. 1 with PS forming the wall and PEO chains stretching from both the inner and outer PS wall surfaces. A close examination of Fig. 3B reveals the existence of undulating nanotubes and the attachment of vesicles to the ends of some nanotubes. These are probably the intermediates of vesicle to tubule transition.

The Eisenberg study led to the following conclusions about nanotube formation. First, it occurred at very low hydrophilic or soluble block contents.

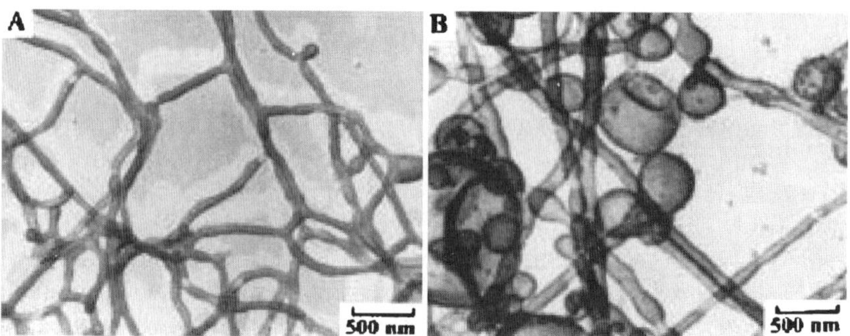

Fig. 3 TEM images of nanotubes (**A**) and vesicle/tube mixtures (**B**) from PS_{240}-PEO_{15}. The aggregates were prepared by equilibrating 2.0 wt. % of the diblock in DMF/water containing 4.5 wt. % of water for 10 (**A**) and 3 (**B**) weeks before addition of water to 25 wt. % and dialysis against water. The images were reproduced with permission from the American Chemical Society

For PS_{240}-PEO_{15}, the PEO weight fraction was only 2.6%. At such a low soluble block content, most other diblock copolymers would have trouble dispersing well in a block-selective solvent. It is probably for this reason that nanotubes have been rarely reported in studies of coil–coil diblock copolymers. Second, the nanotubes formed probably via the fusion of vesicles. The fusion was possible because of the low water content and thus sufficient chain mobility during the aggregate annealing stage in the initial DMF/water mixture. The short PEO block facilitated vesicle fusion for the weak steric stabilization [41] that it rendered to the vesicles.

While PEO was a semi-crystalline polymer, it behaved essentially as random coils in a solvated state in water. Thus, the Eisenberg system fell under the category of coil–coil diblock copolymers. Another example of this category was nanotube formation from a family of coil–coil–coil ABA triblock copolymers poly(2-methyloxazoline)-*block*-poly(dimethylsiloxane)-*block*-poly(2-methyloxazoline)or PMOXA-PDMS-PMOXA, where the PMOXA blocks were water soluble.

PMOXA-PDMS-PMOXA

In their systematic study, Meier and coworkers [17] employed five triblock copolymers with n varying between 62 and 72 and m varying between 11 and 21. These corresponded to the PDMS-to-PMOXA weight ratios from 1.5 to 2.4. To prepare the nanotubes, an effective protocol was to dissolve the copolymer in chloroform, evaporate chloroform in a test tube under rotation, and then to stir the resultant film with water. This method yielded tubes with PDMS making up the wall and PMOXA stretching from the inner and outer wall surfaces. They observed that nanotubes were formed only when the PDMS to PMOXA weight ratio was above 1.5. In aggregates formed from all of the five triblock copolymers, vesicles and nanotubes co-existed. As the PDMS to PMOXA weight ratio was increased, the nanotube population increased.

Nanoaggregate morphological transitions normally occur in the order from spheres to cylinders and then vesicles. Vesicles are formed usually at fairly low soluble block weight fractions. The Meier study [17] indicated that nanotubes were formed at an even lower soluble block weight fraction, a conclusion that seems to be in agreement with the observations by Eisenberg and coworkers. However, for the ABA triblock copolymers, the size asymmetry between the A and B blocks was much smaller than that for the PS-PEO diblock copolymers. Even at a PDMS and PMOXA weight ratio of 2.4, the weight fraction of PMOXA was still high at 29%. Thus, the use of ABA triblock copolymers offers probably a better way to prepare nanotubes for practical applications because one should be able to load more of a foreign agent into nanotubes with longer coronal chains before they start to agglomerate and precipitate.

Aside from the coil–coil diblocks and coil–coil–coil ABA triblocks, crystalline–coil poly(ferrocenyldimethylsilane)-*block*-poly(dimethyl siloxane) or PFS-PDMS diblock copolymers were reported by Raez, Manners, and Winnik [42] to form nanotubes readily in block-selective solvents hexane and *n*-decane, which solubilized the rubbery PDMS blocks and not the crystalline PFS blocks.

PFS-PDMS

Figure 4 shows TEM images of nanotubes prepared from PFS_{40}-$PDMS_{480}$ and PFS_{80}-$PDMS_{960}$. The light central regions and the sandwiching dark lines of these ribbon-like structures were assigned as the cavities and the PFS walls of the nanotubes. The PDMS chains were not visible here and should stretch from both the inner and outer surfaces of the PFS walls.

The fact that the PFS-PDMS diblocks at a coronal to core chain repeat unit number ratio of 12 formed elongated tubular nanoaggregates was surprising, because a coil–coil diblock at this composition would have formed spherical micelles. Such an abnormal behavior has been observed also for another PFS-PDMS polymer [43] with a PFS-to-PDMS repeat unit ratio of 1/18 and for

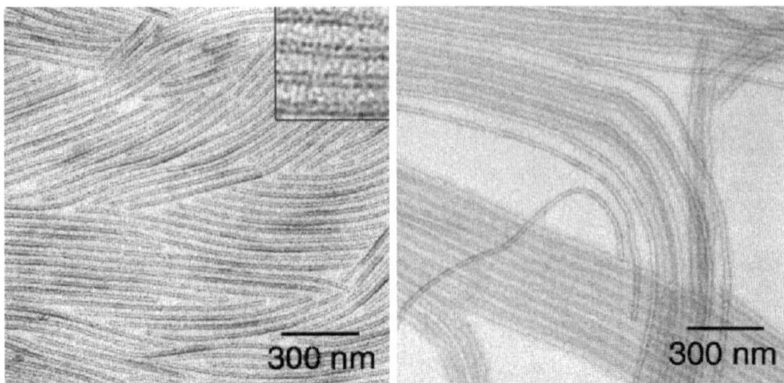

Fig. 4 TEM images of PFS$_{40}$-PDMS$_{480}$ (*left*) and PFS$_{80}$-PDMS$_{960}$ (*right*) nanotubes prepared by dispersing the diblocks in *n*-decane at 61 °C and then cooling at 23 °C for 2 and 24 h, respectively. The images were reproduced with permission from the American Chemical Society

PFS$_{48}$-PMVS$_{300}$ [44], where PMVS denotes poly(methylvinylsiloxane) and is cross-linkable. Such a behavior has been attributed to the crystallinity of the PFS [45] and may occur for other crystalline–coil diblock copolymers as well.

According to Vilgis and Halperin [46], the different micellization behavior was caused mainly by the difference in the free energy F_c per core chain in a crystalline vs. an amorphous core. For an amorphous core, F_c is the energy required to stretch a single core chain. The free energy of a chain in a crystalline core changes depending on how it folds, and F_c increases as it makes folds. The more the core chain folds, the thinner the micellar core is, and the more spaced out the coronal chains are. Thus, the core chain folding helps reduce coronal chain repulsion but increase F_c and the interfacial area between the core and solvent. In the Vilgis and Halperin derivation [46], F_c was combined into the interfacial free energy term for the fold surfaces. The exact morphology and size of micelles formed are those that help minimize the total free energy of the system. For the different energetic expressions or considerations, it is thus not surprising that the morphological transitions should occur at different copolymer compositions for micelles of the two types of diblock copolymers.

The Vilgis and Halperin theory [46] does help shed light on the possible different morphological behaviors of micelles of coil–coil and crystalline–coil diblock copolymers. It can not be used to predict quantitatively the copolymer compositions at which the different micellar morphological transitions take place, because only scaling relations and not quantitative relations were derived for the free energies of three types of micelles. Aside from the semi-quantitative nature of the free energy expressions, the theory did not discuss tubular micelles at all. It examined only hairy disks, star-like micelles with a cubic core, and cylindrical micelles with a cubic-prism-shaped core.

The last reported diblock copolymer family that formed tubular aggregates in block-selective solvents was poly(phenylquinoline)-*block*-polystyrene or PPQ-PS, where PPQ was a rigid-rod block [47]. Such tubes are not discussed further for the following reasons: First, the tubes had diameters of several micrometers and were not nanotubes. Second, the formation mechanism and chain packing in such tubes were not well understood at all. While Halperin [48] has developed a scaling theory for micelle formation from rod-coil diblock copolymers with the rod block forming the core, the theory did not apply to the PPQ-PS system as the block-selective solvents used were good for the rod PPQ block rather than the coil PS block.

Before the conclusion of this section, it should be mentioned that the stability of the self-assembled nanotubes under limited conditions may restrict their applications. To overcome this, Meier and coworkers [17] as well as Winnik, Manners, and coworkers [44] performed crosslinking reactions to their self-assembled nanotubes. The latter team then used their cross-linked nanotubes as template for the preparation of Ag nanoparticles encapsulated in solvent-dispersible nanotubes [49].

In summary, block copolymer tubular micelles or tubular aggregates are not a well-understood subject. To the best of my knowledge, there have been no theories so far examining the formation and properties of tubular micelles. Experimental results have thus far been rare and further experimental studies are required to establish the optimal polymer compositions and other experimental parameters that effect nanotube formation.

3
Cross-linked ABC Triblock Copolymer Nanotubes

3.1
Preparation

There have been only few reports on nanotube formation from the self-assembly in block-selective solvents of copolymers consisting of only coiled blocks. Since nature abhors vacuum, the spontaneous formation of tubular structures from block copolymers in bulk has not been reported and is probably impossible. While the direct preparation of block copolymer nanotubes by self-assembly has been so far difficult, it has been relatively easy to prepare cylindrical nanoaggregates or micelles from ABC triblock copolymers in selective solvents for A only. In such aggregates or micelles, the A block comprises the corona and the C and B blocks comprise the core/shell cylinders. In bulk at the right triblock copolymer composition, the different blocks of an ABC triblock copolymer segregate predictably into C and B core/shell cylinders dispersed in the A matrix [50, 51], if the interfacial tension between the A and C blocks are comparable to that between the A and B blocks and that between the B and C blocks. My group has taken advantage of these observations and chemically processed these precursors to nanotubes. Reviewed in this sub-section is the preparation of nanotubes derived from ABC triblock copolymers.

3.1.1
Nanotube Preparation from Micellar Precursors

The first ABC triblock copolymer used by us to prepare nanotubes was poly*isoprene*-*block*-poly(2-cinnamoyloxyethyl methacrylate)-*block*-poly(*tert*-butyl acrylate) or PI-PCEMA-PtBA [19]:

PI-PCEMA-PtBA

Two samples were used in that study. One was PI_{130}-$PCEMA_{130}$-$PtBA_{800}$, where the PI block, prepared by anionic polymerization in cyclohexane, consisted of 90% 1,4-addition and 10% 3,4-addition units. Nanotube preparation involved first dispersing the triblock in methanol (MeOH). Since only the PtBA block was soluble in MeOH, the triblock copolymer self-assembled into cylindrical aggregates consisting of a PtBA corona and a PI core encapsulated by an insoluble PCEMA intermediate layer (A → B, Fig. 5). The aggregates in MeOH were then subjected to UV light irradiation to cross-link the PCEMA pendant double bonds via 1,4-cycloaddition (B → C, Fig. 5). Nanotubes with hollow centers were obtained after the degradation or the "sculpturing away" of the PI core by ozonolysis and fragment extraction using a solvent (C → D, Fig. 5).

The removal of the PI block was demonstrated by infrared absorption and TEM analyses. More importantly, rhodamine B could be loaded into the tubu-

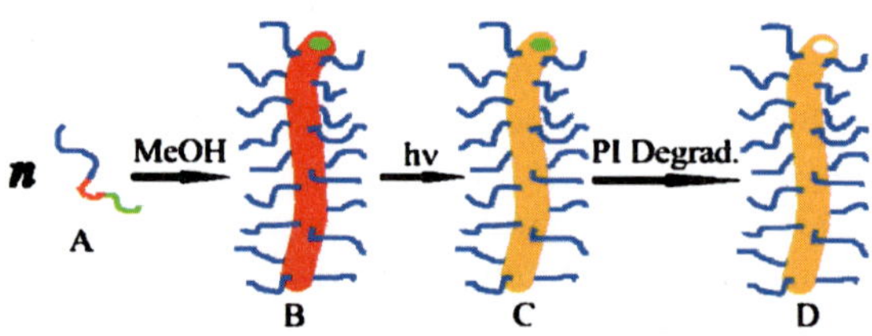

Fig. 5 Steps invoked to prepare nanotubes from PI-PCEMA-PtBA

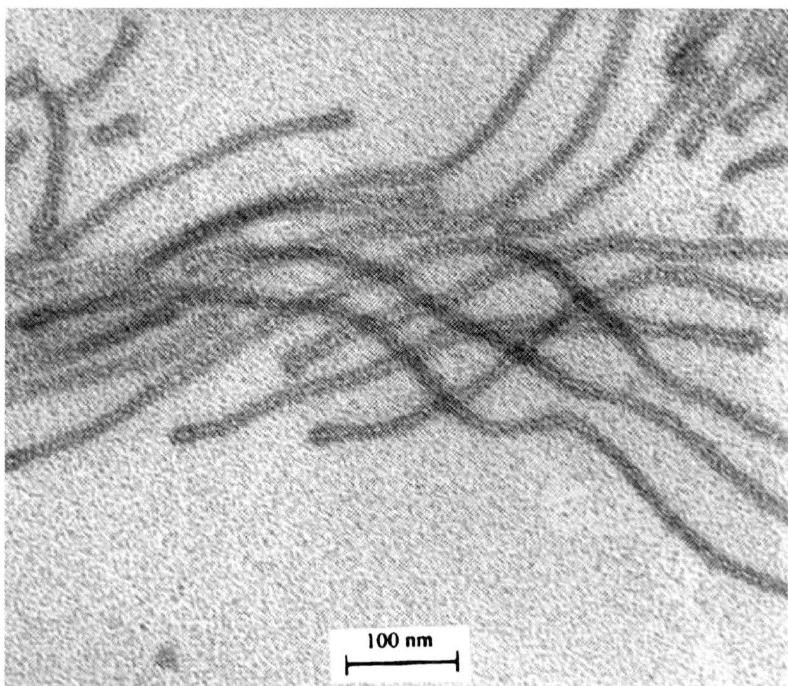

Fig. 6 TEM image of PCEMA-P*t*BA nanotubes prepared from degradation of the PI core of PI-PCEMA-PtBA nanofibers

lar core. Figure 6 shows a TEM image of such nanotubes stained by OsO_4. The center of each tube appears lighter than the PCEMA intermediate layer because the PI block was removed.

The cylindrical nanoaggregate precursor approach has been used to prepare nanotubes from other triblock copolymers including PGMA-PCEMA-PtBA, where PGMA denotes poly(glyceryl methacrylate) [52].

$$\left.\left[CH_2\!-\!\underset{\underset{\displaystyle O}{\overset{\displaystyle CO}{|}}}{\overset{\overset{\displaystyle CH_3}{|}}{C}}\right.\right]_n \qquad \textbf{PGMA}$$

To prepare nanotubes from $PGMA_{375}$-$PCEMA_{120}$-$PtBA_{120}$, the triblock was stirred for 3 to 5 days with water, which was selective for PGMA, to effect formation of cylindrical nanoaggregates first. Such aggregates possessed a PtBA core, a PCEMA shell, and a PGMA corona. Nanofibers were obtained again by photocrosslinking the PCEMA shell. Nanotubes with PAA-lined cores were obtained after the hydrolysis of the PtBA core chains.

Cylindrical nanoaggregates were prepared from tetrablock copolymer PI_{185}-$PtBA_{15}$-$P(CEMA_{67\%}HEMA_{33\%})_{85}$-$PGMA_{245}$ analogously by stirring the copolymer in water. Here $P(CEMA_{67\%}$-$HEMA_{33\%})$ denotes a random block consisting of 67% CEMA and 33% of 2-hydroxyethyl methacrylate [53]. Such nanoaggregates had a cylindrical core consisting of concentric PI, PtBA, and P(CEMA-HEMA) layers. The crosslinking of the PCEMA shell then led to nanofiber formation. Nanotubes were obtained by degrading the PI block by ozonolysis and fragment extraction.

While nanotube formation from the self-assembly of block copolymers required low weight fractions of the soluble coronal blocks, this requirement appears to be much relaxed for the preparation of cylindrical nanoaggregates from ABC triblock copolymers with a solid C core. For PI_{130}-$PCEMA_{130}$-$PtBA_{800}$, the weight fraction for the MeOH-soluble PtBA block was 74%. The PGMA weight fractions in $PGMA_{375}$-$PCEMA_{120}$-$PtBA_{120}$ and PI_{185}-$PtBA_{15}$-$P(CEMA_{67\%}HEMA_{33\%})_{85}$-$PGMA_{245}$ were 56% and 44%, respectively. These numbers are far larger than 2.6% reported by Eisenberg and coworkers [15, 16] for the PEO block of PS_{240}-PEO_{15} for nanotube formation, and thus the nanotubes here should be far more stable in their preparation solvents.

It should be pointed out that the morphologies of nanoaggregates and micelles of ABC triblocks [54, 55] have not been well studied despite the fact that there have been many reports on morphologies of AB diblocks [25]. It will be interesting to investigate if nanotubes can be directly obtained from the self-assembly of ABC triblock copolymers in selective solvents. One promising strategy may be to prepare micelles from ABC triblock copolymers with the A and C blocks highly incompatible in a selective solvent for A and C. If A and C are moderately incompatible, A and C may segregate into patched A and C domains sharing the same coronal surface [56, 57]. The high incompatibility of A and C may require the A and C blocks to segregate on to different coronal faces. These can be the outer and inner surfaces of a nanotube or a vesicle [58] with its wall made of B.

3.1.2
Nanotube Preparation from Solid Precursors

In the absence of strong interactions such as H-bonding and electrostatic interactions, different polymers do not mix. Most block copolymers undergo block segregation or microphase separation in the solid state [51]. If the block copolymers have narrow molecular weight distributions and homogenous composition, highly periodic structures are formed. For ABC triblock copolymers, there are compositions at which periodic B-C shell–core cylinders are

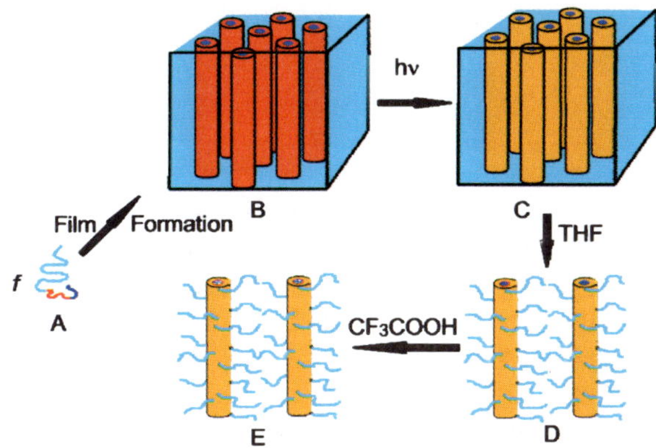

Fig. 7 Schematic illustration of processes involved in the preparation of PS-PCEMA-PAA nanotubes

formed in an A matrix. One can use such structures as precursors for the preparation of nanotubes.

Figure 7 depicts how we prepared PS-PCEMA-PAA nanotubes from PS_{690}-$PCEMA_{170}$-$PtBA_{200}$ via the solid-state precursor approach [20]. Step 1 (A $\rightarrow$ B in Fig. 7) involved casting films from a solution mixture of the triblock copolymer and some PS homopolymer (hPS) in toluene. Some hPS was added to increase the PS volume fraction to $\sim$ 70%. At the volume fractions of $\sim$ 10% and $\sim$ 20% for PtBA and PCEMA, the two blocks should form core-shell cylinders dispersed in the PS matrix [50]. In step 2 (B $\rightarrow$ C), the block-segregated copolymer film was irradiated with UV light to cross-link the PCEMA shell cylinder. The cross-linked cylinders were levitated from the film in step 3 by stirring the film in THF (C $\rightarrow$ D). PS-PCEMA nanotubes containing PAA-lined tubular cores were obtained after removing the t-butyl groups of the PtBA block with trifluoroacetic acid in methylene chloride (D $\rightarrow$ E).

The left panel of Fig. 8 shows a TEM image of a thin section of a hPS/PS-PCEMA-PtBA film. On the right side of this image we see many concentric light and dark ellipses with short stems. These represent projections of cylinders with PCEMA shells and PtBA cores aligned slightly off the normal direction of the image. The PCEMA shells appear darker, because OsO_4 stained PCEMA selectively. The diameter of the PtBA core is $\sim$ 20 nm. On the left side of this image we see cylinders lying in the image plane. Thus, the orientation of the cylindrical domains varied from one grain of the size of micrometers to another for the fact that no special measures [59, 60] were taken to effect their macroscopic alignment.

The right panel of Fig. 8 shows a TEM image of the intestine-like nanotubes. The stained PCEMA layer does not have a uniform diameter across the nanotube length because of their uneven collapse during solvent evaporation.

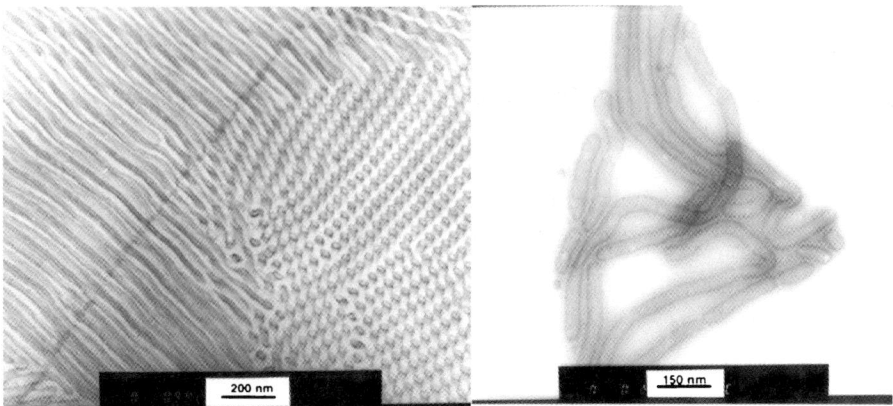

Fig. 8 (*Left*) TEM image of a thin section of the PS_{690}-$PCEMA_{170}$-$PtBA_{200}$/hPS solid. (Right) TEM image of the PS-PCEMA-PAA nanotubes

The presence of PAA groups inside the tubular core was demonstrated by our ability to carry out various aqueous reactions inside the tubular core as will be discussed later.

The solid-state-precursor approach yields nanotubes with a very thick corona. If the solid film was annealed long enough to allow improved block segregation, the precursory nanofibers and therefore nanotubes produced could be very long, e.g., up to hundreds of micrometers long. Two drawbacks of the solid-state-precursor approach are the low yield and the difficulty in obtaining uniform crosslinking among different tubes or even in different sections of the same tube. Uniform crosslinking of PCEMA was difficult to achieve in a PS-PCEMA-PtBA film because light has a finite penetration depth into a film and the surface layers may get more heavily cross-linked than the inner layers. Nanofibers could be obtained at a yield of $\sim 50\%$ because the complete disentanglement of long nanofibers even in a good solvent for the matrix PS block after PCEMA crosslinking was difficult. This is in stark contrast to the quantitative yield that can be achieved in the micelle-precursor approach.

3.2
Dilute Solution Properties

As one-dimensional objects, cylindrical micelles and polymer nanotubes have many features in common with semi-flexible polymer chains, but on a different size scale. Nanotubes tend to be longer, thicker, and more rigid than individual polymer molecules, but both are characterized by a distribution of end-to-end lengths, a radius of gyration, and a persistence length. Figure 9 compares the structures of a poly(*n*-hexyl isocyanate) or PHIC chain, a PS-

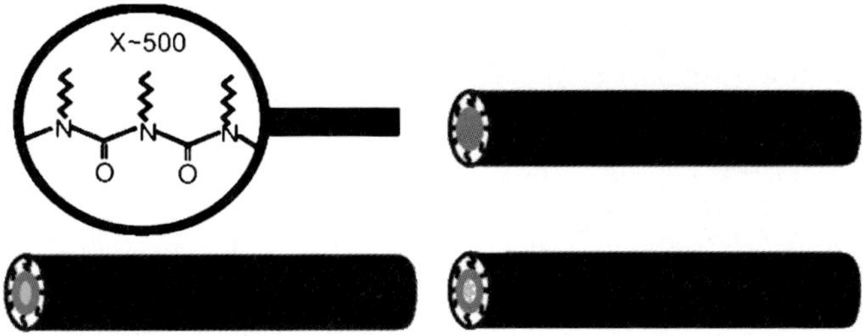

Fig. 9 Structural comparison of a PHIC chain under a "magnifying glass" (*top left*), a PS-PCEMA nanofiber (*top right*), a PS-PCEMA-PtBA nanofiber (*bottom left*), and a PS-PCEMA nanotube with a PAA-lined core (*bottom right*)

PCEMA nanofiber, a PS-PCEMA-PtBA nanofiber, and a PS-PCEMA nanotube with a PAA-lined core.

In a PHIC chain, the backbone, covered by a corona of hexyl groups, consists of a linear sequence of imide units. Their counterparts in the PS-PCEMA nanofiber are the cross-linked PCEMA cylindrical core and the PS chains as the corona. Other than the size difference, the PHIC molecule and the PS-PCEMA nanofiber bear a remarkable structural resemblance. If the PCEMA crosslinking density is high, the PtBA chains in a PS-PCEMA-PtBA nanofiber and the PAA chains in a PS-PCEMA-PAA nanotube become trapped inside the cores even if the solvent is good for both PtBA and PAA. Thus, the triblock copolymer nanofibers and nanotubes can be viewed as giant polymer chains as well [18].

We have so far developed techniques for the fractionation and characterization of diblock copolymer nanofibers and compared the viscosity properties of dilute solutions of diblock copolymer nanofibers and worm-like polymer chains. Based on our structural analysis above, we believe that these techniques and conclusions should apply equally well to triblock nanofibers and nanotubes, and these topics are thus briefly reviewed below.

While we have prepared nanofibers from several families of block copolymers [18, 61–63], the discussion here will be restricted to PS-PI nanofibers. These fibers were prepared by first dispersing PS_{130}-PI_{370} in N,N-dimethyl acetamide, a selective solvent for PS, to effect cylindrical micelle formation. The PI cores were then cross-linked by adding sulfur monochloride S_2Cl_2 [64]. This reagent cross-linked the PI via the following reaction:

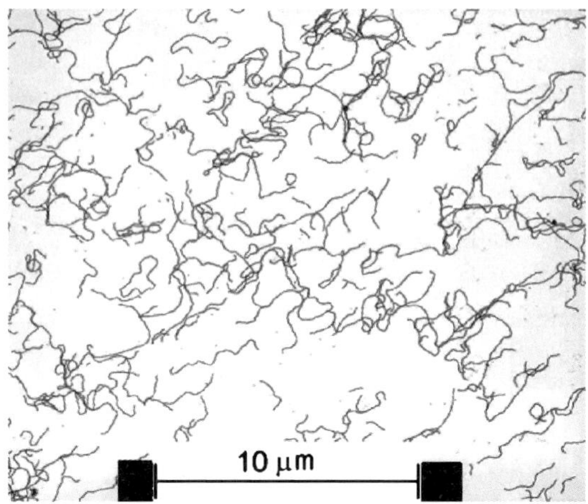

Fig. 10 TEM image of PS_{130}-PI_{370} nanofiber fraction 1

Figure 10 shows a TEM image of the nanofibers thus prepared after aspiration onto a carbon-coated copper grid from THF. The lengths of more than 500 fibers were measured manually for this sample. Figure 11 shows the length distribution function of this sample determined this way. This sample is denoted as fraction 1 or F1 in the figure. From the length distribution function, the weight- and number-average lengths L_w and L_n were obtained. The L_w and polydispersity L_w/L_n valued were 3490 nm and 1.35, respectively, for this sample.

It is difficult to study the dilute solution properties of very long nanofibers. They tend to sediment, and this makes quantitative measurements difficult. While ultracentrifugation [18] or density gradient centrifugation could have

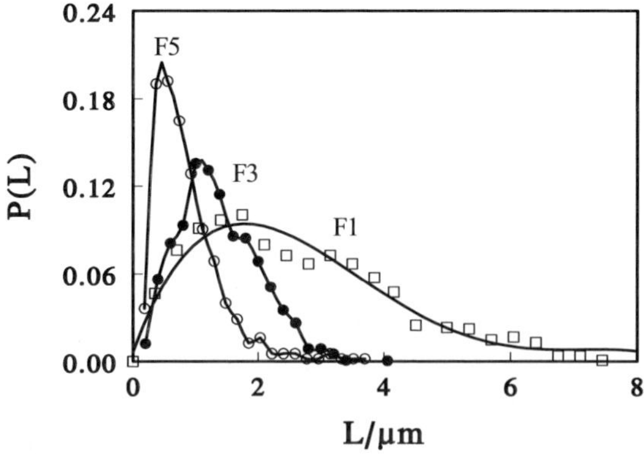

Fig. 11 Plot of fiber population density $P(L)$ vs. length L for PS_{130}-PI_{370} nanofiber fractions 1 (□), 3 (●), and 5 (○) generated from TEM image analysis

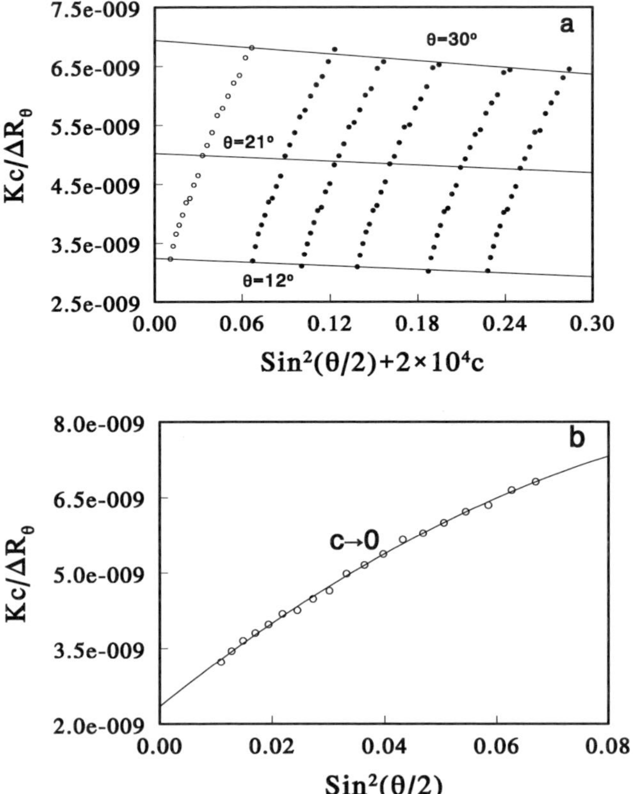

Fig. 12 Zimm plot for the light scattering data of F3 in the scattering angle range of 12 to 30°. The *solid circles* represent the experimental data. The *hollow circles* represent the extrapolated $Kc/\Delta R_\theta|_{c\to0}$ data. In **a**, linear extrapolation of data to zero concentration at the highest and lowest scattering angles of 30 and 12° is illustrated. **b** shows the result of curve fitting of the $Kc/\Delta R_\theta|_{c\to0}$ data fitting by Eq. 1

been used to separate F1 further into fractions with longer and shorter average lengths, we chose to prepare nanofiber fractions with shorter lengths by breaking up the F1 nanofibers by ultrasonication [64] By adjusting the ultrasonication time, we produced fibers of different lengths. In Fig. 11 we present also the length distributions for the samples denoted as F3 and F5 that were ultrasonicated for 4 and 20 h, respectively. As ultrasonication time increased, the distribution shifted to shorter lengths.

These fiber fractions were sufficiently short so that we could determine their weight-average molar masses M_w by light scattering. Figure 12 shows a Zimm plot for the light scattering (LS) data of sample F3 in the scattering angle θ range of 12 to 30°. The data quality appears good. Multiple runs of the same sample indicated that the data precision was high.

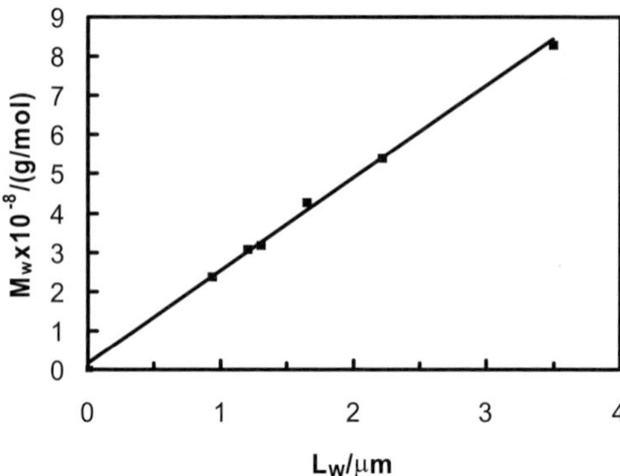

Fig. 13 Increase in LS M_w with TEM L_w for PS_{130}-PI_{370} nanofiber fractions

For the large size of the fibers, the $Kc/\Delta R_\theta$ data varied with $\sin^2(\theta/2)$ or the square of the scattering wave vector q non-linearly despite the low angles used. We fitted the data using

$$\frac{Kc}{\Delta R_\theta} = \frac{1}{M_w}\left[1 + (1/3)q^2R_G^2 - kq^4R_G^4\right] + 2A_2c, \tag{1}$$

and obtained M_w, radii of gyration R_G, and second virial coefficient A_2 for the different fractions. Figure 13 plots the resultant M_w vs. L_w, where L_w were obtained from TEM length distribution functions $P(L)$. The linear increase in M_w with L_w suggests the validity of the M_w values determined.

We measured the viscosities of dilute solutions of the nanofiber fractions in THF using a home-built rotating cylinder viscometer [65] with a very low shear rate γ of 0.082 s^{-1}. Such a low shear rate was used because we found that the viscosity of a nanofiber solution was constant only below $\gamma \approx 0.1$ s^{-1}, and it decreased as the shear rate increased beyond $\gamma \approx 0.1$ s^{-1} [61]. Thus, such solutions were shear thinning just like polymer solutions. Shear thinning happened probably for the alignment of the nanofibers along the shearing direction. Polymers of ordinary molar masses, e.g., < 10^6 g/mol, experience shear thinning only if $\gamma > 10^4$ s^{-1} [66]. The large difference in the γ values of transition from the Newtonian to non-Newtonian flow should be a direct consequence of the drastically different sizes of polymer chains and nanofibers.

The viscosities thus obtained for four nanofiber fractions in THF are plotted in Fig. 14 in the form of a $(\eta_r - 1)/c$ vs. c + LZ plot. Here, c is the nanofiber concentrations c, and η_r is the relative viscosity, which is defined as the ratio between the viscosities of the nanofiber solution and solvent THF. The solid lines represent the best fit to the experimental data by:

$$(\eta_r - 1)/c = [\eta] + k_h [\eta]^2 c, \tag{2}$$

where $[\eta]$ is the intrinsic viscosity and k_h is the Huggins coefficient. The linear dependence between $(\eta_r - 1)/c$ and c is in striking agreement with the

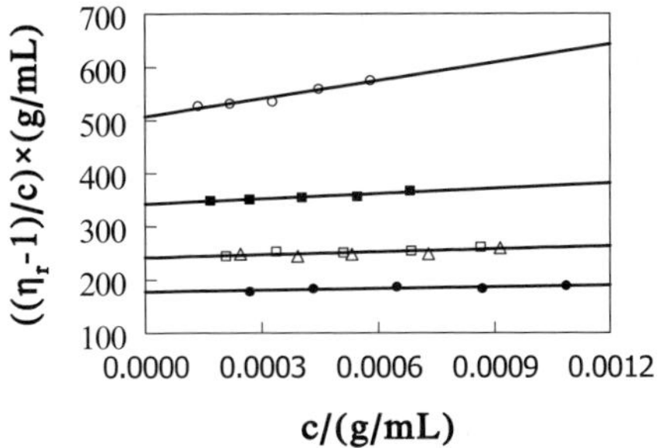

Fig. 14 From *top* to *bottom*, plot of $(\eta_r - 1)/c$ vs. c for PS_{130}-PI_{370} nanofiber fractions 2, 3, 4, and 6 in THF. All the η_r data were obtained using the viscometer at a shear rate of 0.082 s^{-1} with exception to those denoted by (Δ) that were obtained at a shear rate of 0.047 s^{-1}

behavior of polymer solutions. Even more interesting, k_h took values mostly between 0.20 and 0.60 in agreement with those found for polymers [66].

We further analyzed the $[\eta]$ data in terms of the Yamakawa–Fujii–Yoshizaki (YFY) theory originally developed for worm-like chains [67, 68]. According to Bohdanecky [69], the YFY theory could be cast in a much simpler form

$$\left(M_w^2/[\eta]\right)^{1/3} = A + BM_w^{1/2} , \tag{3}$$

for chains in a wide range of reduced chain lengths. In Eq. 3, A and B are fitting parameters that are related to the persistence length l_P and the hydrodynamic diameter d_h of the chains, respectively. Figure 15 shows the data, that we obtained for solutions of PS_{130}-PI_{370} nanofibers in THF plotted according to Eq. 3. From the intercept A and slope B of the straight line, we calculated l_P and d_h for the nanofibers to be (1040 ± 150) and (69 ± 18) nm, respectively.

This procedure was repeated for the nanofibers in different solvents. Table 1 summarizes the l_P and d_h values that we determined in three different solvents for the PS_{130}-PI_{370} nanofibers. The d_h value in THF compares well

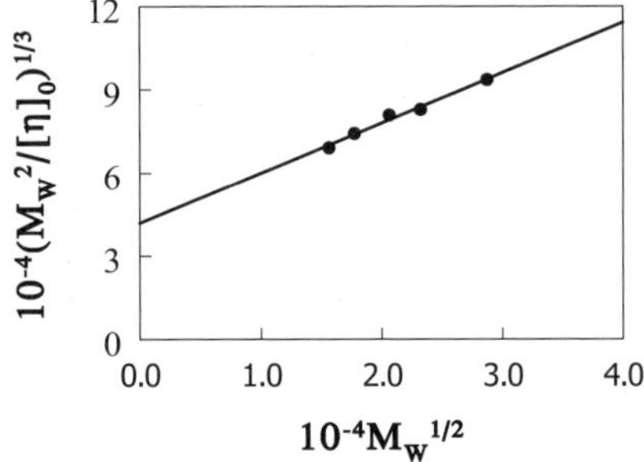

Fig. 15 Nanofiber viscosity data plotted following the Bohdanecky method

Table 1 Persistence length l_P and hydrodynamic diameter d_h of the nanofibers calculated from the viscosity data of PS_{130}-PI_{370} nanofiber fractions in different solvents

Solvent	d_h/nm	l_P/nm
THF	69 ± 18	1040 ± 150
THF/DMF = 50/50	61 ± 18	850 ± 90
THF/DMF = 30/70	51 ± 12	830 ± 60

with what we estimated from the sum of the diameters of the cross-linked PI core, determined from TEM images, and the calculated root-mean-square end-to-end distance of the PS coronal chains. This result supports our idea of the applicability of the FYF theory to the nanofiber solutions. What is more convincing is the trend for decreasing fitted values of d_h values with increasing DMF content in THF/DMF mixtures. While both THF and DMF solubilize PS, d_h decreased with increasing DMF content because the extent of swelling for the cross-linked PI core decreased with increasing DMF content.

In Table 1 we present the l_P values determined in THF and two different THF/DMF mixtures. These values, on the order of 1 µm, are comparable to those reported by Discher and coworkers [38, 70] and by Bates and coworkers [71, 72] for PEO-PI cylindrical micelles with a core diameter of ~ 20 nm in water. Here PEO denotes poly(ethylene oxide). Bates and coworkers deduced their values of l_P from small-angle neutron scattering experiments, whereas Discher and coworkers determined the l_P values using fluorescence microscopy. The fact that the l_P values that we determined from viscometry are comparable to those of the PEO-PI cylindrical micelles with similar core diameters again suggests the validity of the YFY theory in treating the nanofiber viscosity data. This study demonstrates that block copolymer nanofibers have dilute solution properties similar to those of semi-flexible polymer chains.

Another property of semi-flexible polymer chains is the formation of nematic phases in concentrated solution. According to the theories of Onsager [73] and Flory [74], polymer chains with $l_P/d_h > 6$ should form a liquid crystalline phase above a critical concentration. We were able to show the presence of such a liquid crystalline phase by polarized optical microscopy for PS-PCEMA nanofibers dissolved in bromoform at concentrations above ~ 25 wt. % [75] Furthermore, we observed that these liquid crystalline phases disappeared when these solutions were heated to a temperature above a well-defined liquid-crystalline-to-disorder transition temperature, T_{ld}. Such observations suggest nanofibers have concentrated solution properties similar to those of semi-flexible polymers.

For the large size difference between the nanofibers and polymer chains, they are bound to have property differences. The liquid crystalline phases for the PS-PCEMA nanofibers in bromoform were obtained at room temperature only after such solutions were sheared mechanically. The liquid crystalline phase disappeared after the solution was heated above T_{ld}. It did not form spontaneously again upon cycling the sample temperature back to room temperature [75], and was re-formed only after the sample was re-sheared at room temperature. This need for sample shearing is in contrast to the spontaneous liquid crystalline phase formation from concentrated semi-flexible polymer solution [76, 77]. It may be due to the low mobility of the nanofibers and thus their sluggishness in achieving the nematic packing order.

Aside from the different mobility of a polymer chain and a nanofiber, the van der Waals forces between nanofibers should be larger than those between polymer chains. These forces should increase with the size of the nanofibers [78] Above a certain size, the interaction between nanofibers may become so strong that the fibers cluster and precipitate even from a good solvent for the coronal chains. When this happens, the analogy between nanofibers and polymer chains fails.

We recently examined the stability of nanofibers dispersed in THF prepared from PS_{130}-PI_{370} [64]. This particular nanofiber sample had $L_{\mathrm{w}} = 1650$ nm, $L_{\mathrm{w}}/L_{\mathrm{n}} = 1.21$, and $M_{\mathrm{w}} = 4.3 \times 10^{8}$ g/mol, respectively. At a concentration of $\sim 8 \times 10^{-3}$ g/mL and under gentle stirring, no nanofiber settling was detected during 4 days of observation by light scattering. Without stirring, we noticed a 10% decrease in the light scattering intensity of the solution, which corresponded to ~ 10 wt. % settling of the nanofibers in the first 4 days. No noticeable further settling was observed in another 8 days [64]. This suggests that the longer fibers in this sample exceeded the size for settling. The fact that the nanofiber solutions in THF had a negative second virial coefficient from static light scattering suggested that the longer fibers probably first clustered and then settled. The fact that the clustering could be prevented by gentle stirring suggests that only a very shallow attraction potential existed between the fibers.

The length for fiber settling was short for the above sample and was probably at several micrometers. This length depends on many factors. Fibers with a long soluble corona block and a short insoluble core block(s) should have high colloidal stability. Decreasing the core diameter should increase fiber colloidal stability as well.

We have not performed any detailed studies of solution properties of block copolymer nanotubes. For the structural similarities between the two, we expect the nanotubes to have similar viscosity and liquid crystalline properties as the nanofibers.

3.3
Chemical Reactions

In this section we return to the analogy between a polymer molecule and nanofibers and nanotubes. A PI chain can be hydrogenated via "backbone modification" to yield a polyolefin chain. It can also be epoxidized or otherwise modified. Polymer chains can be grafted onto a solid substrate to form a brush layer [79–81]. Through techniques like anionic polymerization one can prepare "end-functionalized" polymers. For the structural similarity between polymer chains and block copolymer nanofibers and nanotubes they should have similar reaction patterns. In this section, we show that nanotubes can undergo backbone modification, surface grafting, and end functionalization reactions.

3.3.1
Backbone Modification

For nanofibers and nanotubes, backbone modification refers to selective chemical transformation of one or more of the core blocks. In our preparation of nanotubes from the PI-PCEMA-PtBA and PS-PCEMA-PtBA triblock nanofibers, the core PI and PtBA blocks were "sculpted" away either fully or partially. These reactions are examples of nanofiber backbone modification. One can also consider inorganic reactions carried out inside one of the block copolymer phases to be a kind of backbone modification. These reactions lead to the formation of interesting polymer–inorganic hybrid nanostructures. This topic will be reviewed in the following paragraphs.

The first report on the preparation of block copolymer/inorganic hybrid nanofibers appeared in 2001. It dealt with the filling of the core of the PS-PCEMA-PAA nanotubes by γ-Fe_2O_3 [20]. The preparation involved first the equilibration between the nanotubes and $FeCl_2$ in THF. Fe(II) entered the nanotube core for their binding with the core carboxyl groups. The extraneous $FeCl_2$ was then removed by precipitating the Fe(II)-containing nanotubes into methanol. Adding NaOH dissolved in THF containing 2 vol. % of water precipitated Fe(II) trapped in the nanotube core as ferrous oxide. The ferrous oxide was subsequently oxidized to γ-Fe_2O_3 via the addition of hydrogen peroxide [82]. The top panel in Fig. 16 shows a TEM image of the hybrid nanofibers, and the γ-Fe_2O_3 particles are seen to be produced exclusively inside the nanotube cores.

The production of γ-Fe_2O_3 in the confined space of the "nano test tubes" resulted in particles that were nanometer-sized. For this, the particles were superparamagnetic as demonstrated by results of our magnetic property measurement [20]. This meant that they were magnetized only in the presence of an external magnetic field and were demagnetized when the field was removed. To see how such fibers behaved in a solvent in a magnetic field, we dispersed the fibers in a solvent mixture consisting of THF, styrene, divinylbenzene, and a free radical initiator AIBN. The fiber dispersion was then dispensed into a NMR tube and mounted in the sample holder of a NMR instrument. In the 4.7-T magnetic field of NMR, the solvent phase was gelled by raising temperature to 70 °C to polymerize styrene and divinylbenzene. Thin

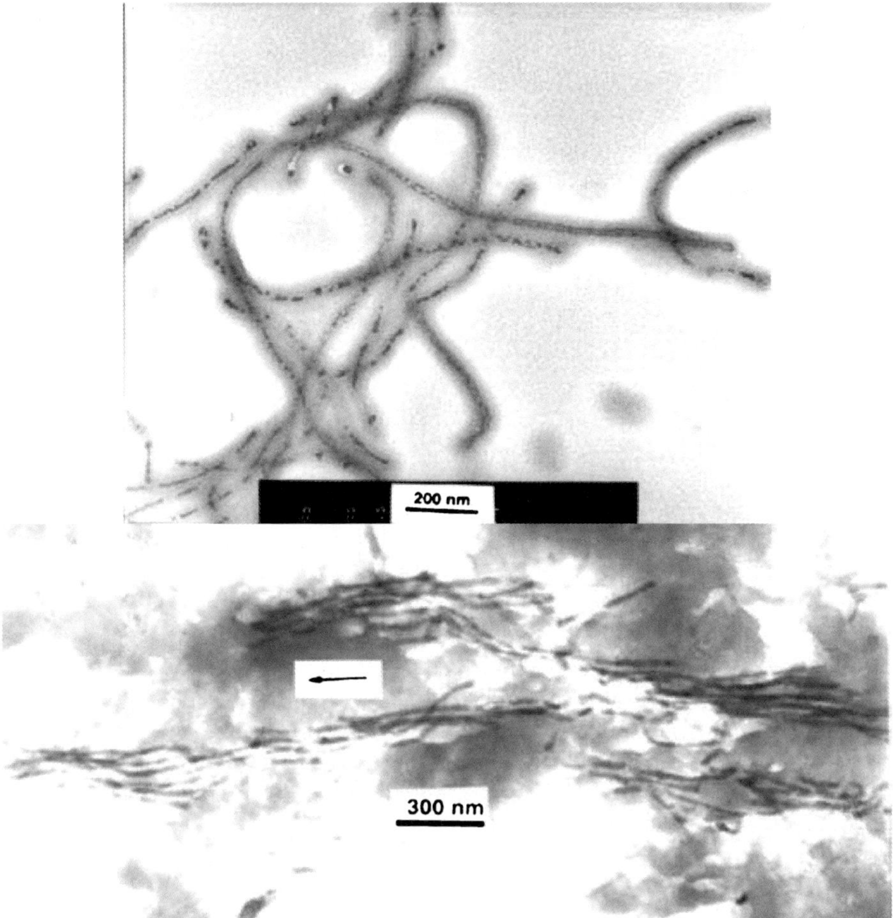

Fig. 16 *Top*: TEM image of PS-PCEMA-PAA/Fe$_2$O$_3$ hybrid nanofibers. *Bottom*: Bundling and alignment of the nanofibers in a magnetic field. The *arrow* indicates the magnetic field direction

sections were obtained from the gelled sample by ultramicrotoming. Shown in the bottom panel of Fig. 16 is a TEM image of nanofibers in a gelled sample. One consequence of the induced magnetization of the fibers is that they attracted one another and bundled in a magnetic field. Also clear from this image is that the fibers aligned along the magnetic field direction.

More recently we derivatized nanotubes to prepare water-dispersible tetra-block/Pd hybrid catalytic nanofibers [53] and triblock/Pd/Ni superparamagnetic nanofibers [83]. To prepare the former, nanotubes were prepared from PI$_{185}$-PtBA$_{15}$-P(CEMA$_{67\%}$-HEMA$_{33\%}$)$_{85}$-PGMA$_{245}$. The hydroxyl groups of the precursor PHEMA block was not fully cinnamated in this tetrablock because the HEMA units facilitated the transport of Pd^{2+} and Ni^{2+} across the nanotube walls. As mentioned before, the nanotube preparation involved first dispersing freshly-prepared PI-PtBA-P(CEMA-HEMA)-PGMA in water to yield cylindrical aggregates consisting of a PGMA corona and a PI core. Sandwiched between these two layers were a thin PtBA layer and a P(CEMA-HEMA) layer. Such cylindrical aggregates were then irradiated to cross-link the (CEMA-HEMA) layer. The PI core was degraded by ozonolysis. By controlling the ozonolysis time, the degree of PI fragmentation could be controlled. When not fully degraded, the residual double bonds of the PI fragments trapped inside the nanotubular core were able to sorb Pd(II) most likely via π-allyl complex formation:

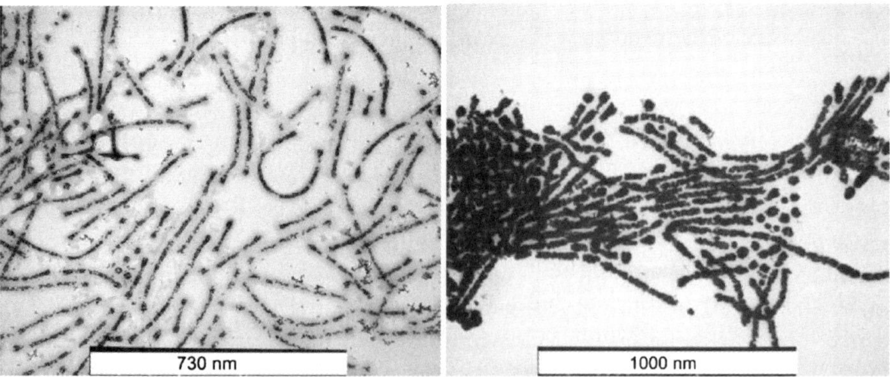

The complexed Pd(II) was then reduced by $NaBH_4$ to Pd. The left panel of Fig. 17 is a TEM image for such nanotubes containing 4.0 wt.% reduced Pd nanoparticles.

The Pd-loaded nanofibers were dispersible in water as well as in water-based electroless plating solutions. Thus, the Pd nanoparticles could be used as a catalyst for the further electroless deposition of other metals. We, for example, plated Pd into the tubular core onto the initially formed Pd nanoparticles via electroless plating to yield essentially continuous Pd nanowires. The right image in Fig. 17 shows a TEM image of such hybrid nanofibers after the incorporation of Pd to a total of 18.4 wt.%. In fact, we could tune the final

Fig. 17 *Left*: TEM image of nanofibers containing 4 wt.% Pd. *Right*: TEM image of nanofibers containing 18 wt.% of Pd. The *scale bars* in the form of white boxes are 730 and 1000 nm long, respectively

amount of Pd loaded into the nanotubes by changing the mass ratio between the original Pd-bearing nanotubes and Pd^{2+} in a plating bath, because the plating reaction was essentially quantitative if enough time was allowed for it to take place. A drawback of high Pd loading, e.g., > 20 wt. %, was the poor dispersibility of the hybrid fibers. In addition to Pd plating, we have also succeeded in plating Ni into the core of such nanotubes. We have, for example, succeeded recently in preparing water-dispersible PGMA-P(CEMA-HEMA)-PAA/Pd/Ni magnetic nanofibers [83].

3.3.2
Surface Grafting

Surface grafting in polymer science refers to the attachment of polymer chains to a solid substrate via reactions between the end groups of polymer chains and a surface. Grafting of polymer chains is important, because it helps change the surface properties of a solid substrate. The grafting of a medical device such as a heart valve by poly(ethylene glycol) helps decrease protein deposition and increase the biocompatibility of the device [84]. The grafting of a polymer on to surfaces of two moving machine parts helps reduce friction between the parts [85]. In the case of inorganic particles, the grafting of polymer chains on their surfaces prevents particle agglomeration and improves particle dispersion in solvent or polymer matrix [41, 86, 87]. Grafted polymers also help with inorganic particle processing and ordering. One can, for example, tune the average distance between neighboring Au nanoparticles in a film of polymer-grafted Au nanoparticles by changing the molecular weight of the polymer [86].

If properly end-functionalized, a block copolymer nanotube may also graft via end group reactions to a solid substrate. The grafting of surfaces with exotic reagents such as block copolymer nanotubes may be more advantageous for certain applications such as the facilitation of cell growth [88] and the creation of superhydrophobic or superhydrophilic surfaces [89, 90]. For the latter application, surface roughness on multi-length scales, i.e. the nanometer and micrometer scale, is required [91]. Surfaces grafted with block copolymer nanotubes may lead also to applications that have not been envisioned before. Reviewed in the next paragraphs is the grafting of PS_{630}-$PCEMA_{125}$-PAA_{135} nanotubes to glass and mica surfaces [92].

Based on the assumption that we could cut short the nanotubes and expose PAA core chains at the nanotube ends, our initial design was to end-graft such nanotubes to surfaces bearing amino groups via the amidization reaction. Such amino groups could be introduced on glass or mica surfaces via silane chemistry well-established for the polymer filler industry. Unfortunately, such a strategy failed. This was probably due to the shielding of the PAA core chains by the much longer PS coronal chains of the nanotubes and the difficulty for the PAA chains to react with the surface amino groups.

We then decided to take a detour to accomplish the nanotube grafting. Such a detour should involve reacting the PAA end groups of the nanotubes with an ABA triblock copolymer. Here the A blocks were polyamine. Since the triblock copolymer was to be used in excess, the copolymer should react with the PAA chains via one A block only. The free polyamine groups of the other A block should now be more accessible than the original PAA chains, because the terminal A block is connected to a nanotube end via a PAA chain and a B block. If the B block is long, the terminal polyamine A block can stretch far beyond the shielding range of the PS coronal chains. The reaction between PAA and ABA should, however, convert the nanotube end groups from PAA to polyamine groups. For surface grafting of these tubes, the mica or glass surfaces should possess carboxyl groups. We planned to obtain such groups by reacting the amino groups on the surfaces of mica and glass with succinic anhydride, PAA, or suberic acid.

We shortened the PS_{630}-$PCEMA_{125}$-PAA_{135} nanotubes by subjecting them to ultrasonication. In one case, the ultrasoniation of a nanotube sample for 8 h reduced the number-average length of the tubes from 518 nm to 187 nm [93] The ABA triblock copolymer used to react with the PAA end chains was $PAES_5$-PS_{200}-$PAES_5$, where PAES denotes poly[4-(2′-aminoethyl)styrene]. To introduce amino groups to the surfaces of mica and glass plates, such substrates were exposed to vapors of aminopropyltrimethoxysilane (APTMS) and *N*,*N*-di*iso*propylethylamine in a dessicator. While the trimethoxylsilane groups of APTMS underwent sol–gel chemistry to form a silica coating bearing amino groups, *N*,*N*-di*iso*propylethylamine functioned as a catalyst for the sol–gel reaction [94]. The surface carboxyl groups were obtained via reacting the amino groups of the APTMS coating with succinic anhydride or PAA.

We found that higher nanotube grafting densities were obtained by reacting the nanotubes with mica or glass surfaces that were last treated by PAA rather than succinic anhydride. Figure 18 shows two AFM images of PS-PCEMA-PAA nanotubes that were grafted to mica surfaces modified first by APTMS and then by PAA [92]. The nanotube grafting density was high. The grafted nanotubes could be classified roughly into lying and standing tubes. The relatively long tubes lay flat on the mica surface. The relatively short tubes stood vertically on the mica surfaces with height typically < 250 nm. The standing tubes were not perfectly cylindrical but assumed the shape of an atomic force microscopic (AFM) tip.

The fact that the nanotube grafting density was higher when PAA rather than succinic anhydride was used to treat the mica plates bearing amino surface groups may have its origin in the accessibility of the carboxyl groups. The treatment of such a mica plate with succinic anhydride produces a monolayer of carboxyl groups on the substrate surface. Such a layer should be concentrated in a thin lateral or essentially 2-d layer. When treated with PAA, the grafted PAA layer might be much thicker. This helped increase the reaction probability between the carboxyl groups and PAES block of the nanotubes.

The standing nanotubes of Fig. 18 assumed the shape of an AFM tip because the AFM tip was not infinitely thin and was imaged by the standing nanotubes. The observation of many standing tubes suggested that the tubes were attached to the mica surfaces mainly via end grafting. The longer tubes lay flat on the mica surface probably because they bent and fell on the mica surface after the solvent that was used to rinse the reacted mica evaporated from mica surface. Most tubes would not have been grafted perfectly vertically on a surface for several possible reasons. For one, the cutting surface of a tube might not be perfectly perpendicular to the tube axial direction, and the grafting of such a tube might slant away from the normal direction. At the same slanting angle, a longer tube should produce a larger down or falling force on the connecting or hinging PAE-PS-PAE chains, and it was thus more likely to fall down after solvent evaporation.

The results above demonstrate the successful grafting of block copolymer nanotubes to mica surfaces that were covered by a uniform layer of PAA groups. Many techniques can be used to chemically pattern a substrate [95]. We decided to pattern glass surfaces into amine-rich and amine-poor regions and then to examine the possibility of achieving the patterned grafting of nanotubes. Figure 19 shows how we achieved the patterned grafting of PS_{630}-$PCEMA_{125}$-PAA_{135} nanotubes.

To obtain a glass plate with an amine-patterned surface, we started by spin-coating a poly(2-cinnamoyloxyethyl acrylate) or PCEA film on the glass plate. The film was then irradiated using a TEM copper grid as the mask to cross-link PCEA in the grid square regions (A → B). PCEA in the unexposed TEM grid bar regions was rinsed off by ultrasonication in chloroform for 2.5 h (B → C). The PCEA-patterned plates were subsequently exposed

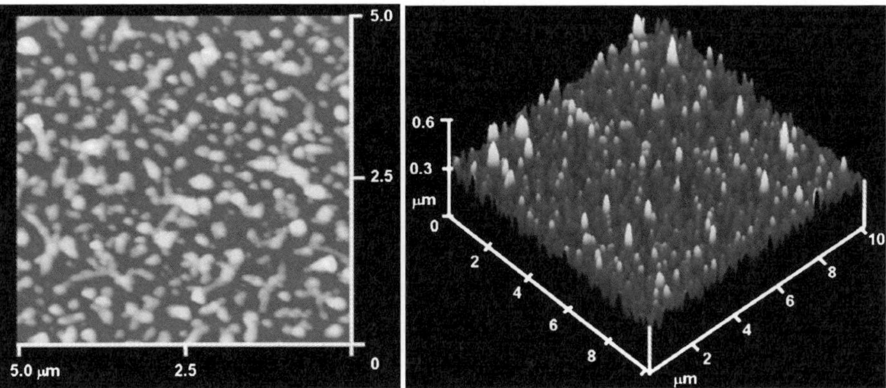

Fig. 18 AFM images of nanotubes grafted on mica surfaces that were modified sequentially by APTMS and PAA

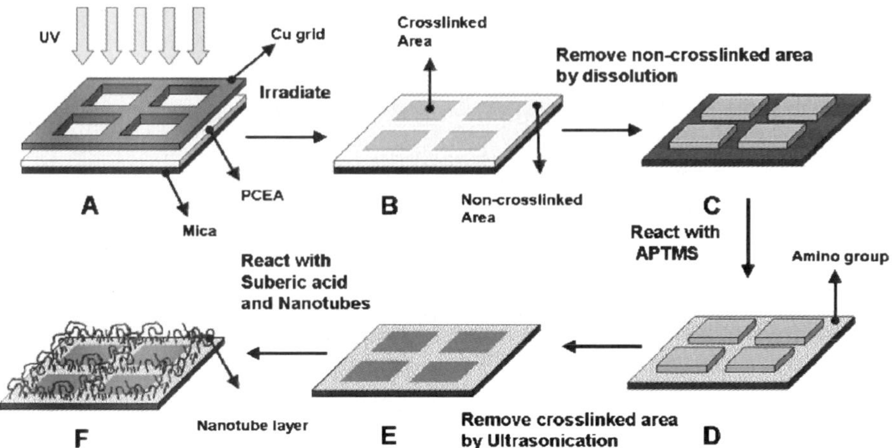

Fig. 19 Processes involved for patterned grafting of PS-PCEMA-PAA nanotubes onto glass plates

to APTMS and DPEA (*N*,*N*-di*iso*propylethylamine, C → D) to grow a gelled APTMS layer in the PCEA free regions. After this treatment, the cross-linked PCEA regions were mostly removed by ultasonication in chloroform for another 3.5 h (D → E). The amino-patterned surfaces were then reacted with nanotubes bearing PAES terminal chains using suberic acid as the connector (E → F) to yield finally nanotube-patterned surfaces.

While steps A → E were self-explanatory, the use of suberic acid in step (E → F) needs some explanation. To graft a monolayer of nanotubes bearing terminal PAES chains to such glass surfaces, we should have followed our prior protocols. That would have involved reacting the amino groups of the glass surfaces with PAA or succinic anhydride and then reacting the newly generated carboxyl groups with the PAES chains. Such a protocol produced a monolayer of grafted nanotubes but was more tedious involving two reaction steps. Suberic acid bearing two terminal carboxyl groups per molecule was used because it could couple the PAES block and the surface amino groups in one step. A drawback of this procedure was that it led also to tube-tube coupling and to the coupling of different glass surface amino groups. Thus, not necessarily a monolayer of nanotubes was grafted this way.

Except steps A → B and C → D, our success in achieving the other steps of Fig. 19 were confirmed by AFM and/or fluorescence microscopy. Fluorescence microscopy could be used, because we were able to attach covalently some fluorescein groups to the PAA core chains of the nanotubes. Figure 20 shows a fluorescence microscopic image of a nanotube-patterned glass surface. The fluorescent nanotubes were expectantly concentrated in the original TEM grid bar regions which were not irradiated during the PCEA film photolysis step. Such a conclusion was confirmed also by AFM experiments with

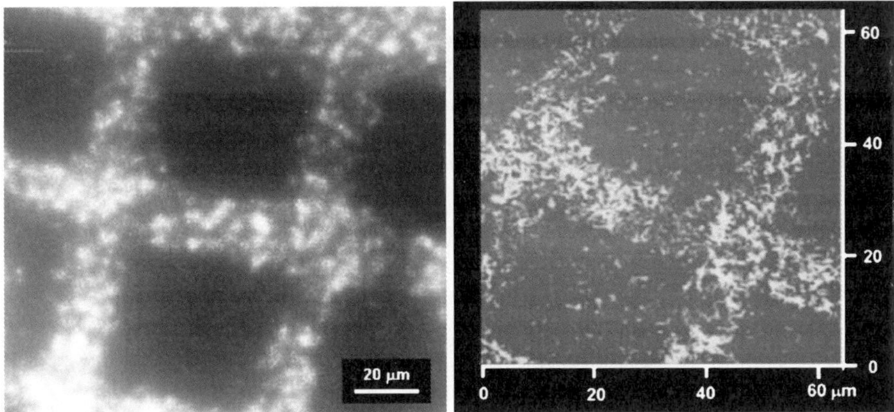

Fig. 20 Fluorescence microscopic (*left*) and AFM (*right*) images of the surface of a glass plate after the patterned grafting of the PS_{630}-$PCEMA_{125}$-PAA_{135} nanotubes

an image also shown in Fig. 20. Aside from confirmation of nanotube concentration in the grid "bar" regions, some nanotubes were seen in the mesh square regions as well. This was probably due to the partial permeability of APTMS and DPEA across the cross-linked PCEA film.

3.3.3
End Functionalization

In the "supramolecular chemistry" of nanotubes, end functionalization should not be interpreted just as the introduction of the traditional functional groups such as carboxyl or amino groups etc. It should also include the introduction of other nano "building blocks" such as nanospheres and nanofibers or nanotubes of a different composition. We first end-functionalized PS_{560}-$PCEMA_{140}$-PAA_{160} nanotubes by attaching to them water-dispersible PAA-PCEMA nanospheres and latex nanospheres prepared from emulsion polymerization to contain surface carboxyl groups [93]. Figure 21 shows the steps involved in coupling the nanotubes and PCEMA-PAA nanospheres.

The first step involved again attaching a triblock copolymer, $PAES_{15}$-PS_{115}-$PAES_{15}$, to the end-exposed PAA core chains of the nanotubes. The resultant nanotubes were then reacted with the carboxyl groups on the surfaces of the nanospheres. The coupling between the nanotubes and emulsion spheres containing surface carboxyl group was achieved similarly.

Figure 22 shows the typical products from coupling the PAES-PS-PAES-treated nanotubes with a batch of emulsion nanospheres bearing surface carboxyl groups.

The product in Fig. 22a resulted from the coupling between one tube and one sphere. Since the spherical "head" is water-dispersible and the tube

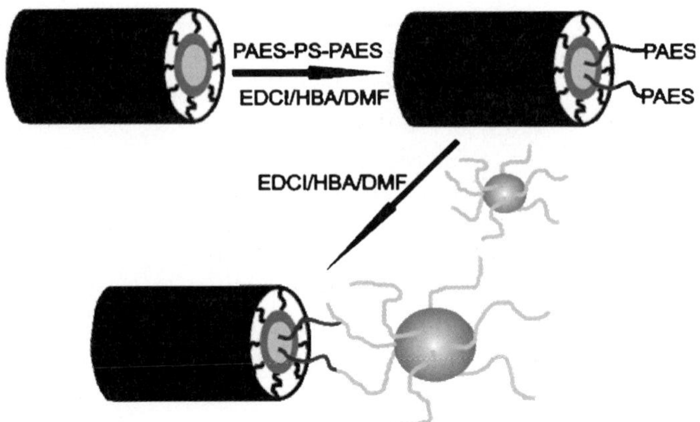

Fig. 21 End functionalization of PS-PCEMA-PAA nanotubes by PAA-PCEMA nanospheres

"tail" is hydrophobic, this structure may be viewed as a macroscopic counterpart of a surfactant molecule or a "super-surfactant". Figure 22b shows the attachment of two tubes to one sphere, which was fused with another sphere probably during solvent evaporation for TEM specimen preparation. "Dumbbell-shaped molecules" were formed by the attachment of two spheres to the opposite ends one tube, as are seen in Fig. 22c. The products shown in Fig. 22a–c co-existed regardless how we changed the tube-to-microsphere mass ratio from 20/1 to 1/20. At the high tube to emulsion sphere mass ratio of 20/1, the super-surfactant and dumbbell-shaped species were the major products. At the mass ratio of 1/1 and 1/20, the dumbbell-shaped product dominated.

Other than product control by adjusting the stoichiometry, an effective method to eliminate the dumbbell-shaped product was to use nanotubes labeled at only one end by PAES-PS-PAES. These tubes were obtained by breaking up ultrasonically nanotubes that contained end-grafted PAES-PS-PAES chains. The reaction between the shortened tubes and the nanospheres at the tube-to-sphere mass ratio of 1/20 yielded almost exclusively the su-

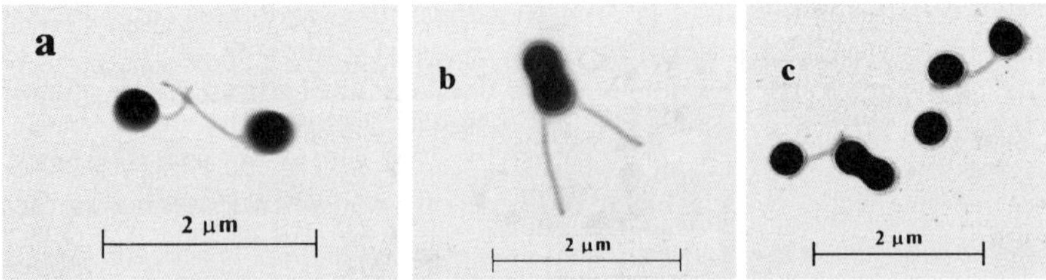

Fig. 22 TEM images of nanotube and emulsion nanosphere coupling products

persurfactant structure containing unreacted nanospheres. The content of the multi-armed structure increased as the nanotube-to-microsphere mass ratio increased.

The same chemistry has been used to couple PS_{560}-$PCEMA_{140}$-PAA_{160} nanotubes with $PGMA_{375}$-$PCEMA_{120}$-PAA_{120} nanotubes [96]. Figure 23 shows the nanotube multiblocks that we prepared. To facilitate the easy differentiation between the two types of nanotubes, we loaded Pd nanoparticles into the PGMA-PCEMA-PAA nanotubes. Using this method, we succeeded in preparing both nanotube di- and tri-blocks with structure similar to di- and tri-block copolymers. The nanotube multiblocks were shown to self-assemble both in a block-selective solvent and in the solid state just as block copolymers but the resultant structures were not very regular yet probably for the poor control in nanotube lengths.

Aside from showing the similarity between the reaction patterns of one-dimensional nanostructures and polymer chains, the coupling methods that we have developed should have other applications. They can, for example, be used to make rudimentary nanomechanical devices. Figure 24 shows, for example, a device that can be prepared by attaching three PS-PCEMA-PAA/γ-Fe_2O_3 hybrid nanofibers to one microsphere.

Being superparamagnetic, these fibers will attract one another in a magnetic field and thus grab a nanoobject. The grabbed nanoobject may be moved around by moving the microsphere through a laser that traps the microsphere by the optical tweezing mechanism. The nanoobject can then be released when desired by turning off the magnetic field. Thus, such a device may function as an "optical magnetic nanohand".

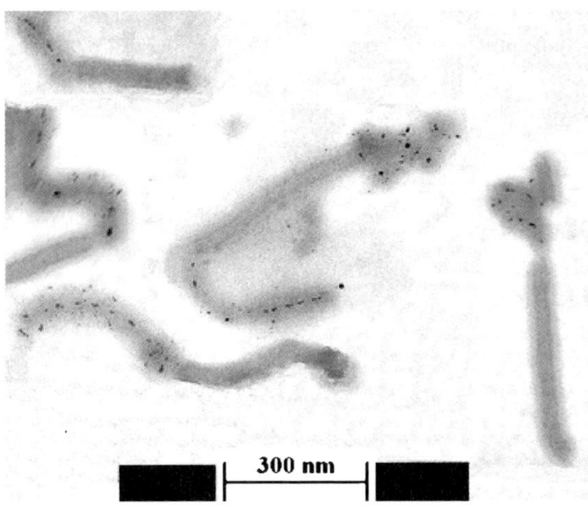

Fig. 23 TEM image of nanotube multiblocks

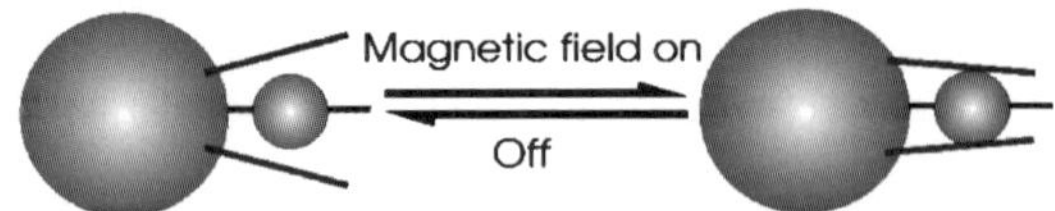

Fig. 24 Schematic illustration of the operation of an optical magnetic nanohand

The coupled nanostructures should be useful for further or tier-II assembly of block copolymers. The super-surfactants of Fig. 22a may, for example, self-assemble like surfactant molecules to form supermicelles with structural order at several length scales. Studies of the self-assembly of the nanotube di- and tri-blocks with structure similar to di- and tri-block copolymers should be of interest as well.

4
Conclusions

Block copolymer nanotubes can be prepared either directly from block copolymer self-assembly in block-selective solvents or from the chemical processing of ABC triblock copolymer nanofibers. There has been only one report on the formation of self-assembled nanotubes from coil–coil AB diblocks in block selective solvents, and it occurred for a sample with a very low weight fraction of the soluble block. Nanotubes were formed from coil–coil–coil ABA triblock copolymers at much higher weight fractions for the soluble A blocks. Still, lower soluble block weight fractions were required for nanotube than for vesicle formation. It remains to be seen if these trends can be generalized to other block copolymers containing purely coil blocks.

The self-assembly of crystalline–coil and rod–coil diblock copolymers in block-selective solvents presented quite some surprises. Crystalline–coil diblocks formed tubular nanoaggregates in block-selective solvents for the coil blocks at coil to crystalline block repeat unit number ratios substantially larger than 1, e.g., 12 and 18 for the PFS-PDMS diblock copolymers. This made the block copolymer nanotubes much easier to access. It again remains to seen if such a trend can be generalized to other diblock copolymers. Thus, much remains to be done to establish the best experimental conditions for formation of self-assembled nanotubes. Theories need to be developed to understand the formation and property of self-assembled block copolymer nanotubes.

Cylindrical nanoaggregates can be prepared from ABC triblock copolymers in solvents selective for one of the end blocks, e.g., A, over a much wider copolymer composition window than nanotubes. They are much easier to access than the nanotubes. Such aggregates consist of an A corona, an insoluble B shell, and an insoluble C core. With a proper copolymer design, one can then crosslink the B shell and sculpt away (either fully or partially) the C core to yield nanotubes. An ABC triblock copolymer undergoes self-assembly also in the bulk or solid state. If the interfacial tension γ_{AC} between A and larger than or comparable to γ_{AB} or γ_{BC}, the C and B blocks segregate in concentric core–shell cylinders dispersed in the A matrix when the volume fractions for B and C are $\sim 20\%$ for B and $\sim 10\%$, respectively. Such a solid morphology can also be chemically processed to yield nanotubes. While the "molecular sculpturing" strategy has so far been characterized by the easiness for finding the cylindrical morphologies with a B shell and a C core either in an A-selective solvent or in the solid state, the additional chemical reaction steps were, however, of inconvenience.

Like small amphiphile nanotubes [14], block copolymer nanotubes should have potential applications in controlled delivery and release [97, 98], in encapsulation [99], and in nanoelectronics [100] etc. Although there have been reports on laboratory use of block copolymer nanofibers as vehicles for drug delivery [101], as scaffolds for cell growth [88, 102], as precursors for ceramic magnetic nanowires [103, 104], and as precursors for carbon nanofibers [105, 106], practical applications of block copolymer nanotubes have not been reported. This is probably due to the difficulty in making such structures. Their preparation from the self-assembly of block copolymers and interests from industry will likely change the scenery of nanotube applications in the future.

Acknowledgements G.L. is very grateful for financial support from the Natural Sciences and Engineering Research Council of Canada, Defence Research and Development Canada, Canada Research Chairs Program, Canada Foundation for Innovations, Ontario Innovation Trust. G.L. would also like to thank Drs. Xiaohu Yan, Zhao Li, Sean Stewart, and Kyoungmoo Koh for performing the experiments described in this article. Prof. M.A. Winnik is gratefully acknowledged for polishing the writing in the manuscript.

References

1. Lee JS, Hirao A, Nakahama S (1989) Macromolecules 22:2602
2. Liu GJ, Ding JF, Guo A, Herfort M, Bazett DJones (1997) Macromolecules 30:1851
3. Liu GJ, Ding JF, Hashimoto T, Kimishima K, Winnik FM, Nigam S (1999) Chem Mater 11:2233
4. Hillmyer MA (2005) Adv Polym Sci 190:137
5. Lazzari M, Liu GJ, Lecommandoux S (2006) Block Copolymers in Nanoscience. Wiley-VCH, Weinheim, Germany
6. Park M, Harrison C, Chaikin PM, Register RA, Adamson DH (1997) Science 276:1401
7. Park C, Yoon J, Thomas EL (2003) Polymer 44:6725
8. Lazzari M, Lopez-Quintela MA (2003) Adv Mater 15:1583
9. Hamley IW (2003) Angew Chem Int Ed 42:1692
10. Thurn-Albrecht T, Schotter J, Kastle CA, Emley N, Shibauchi T, Krusin-Elbaum L, Guarini K, Black CT, Tuominen MT, Russell TP (2000) Science 290:2126
11. Martin CR (1994) Science 266:1961
12. Chen JT, Zhang MF, Yang L, Collins M, Parks J, Avallone A, Russell TP (2007) J Polym Sci B Polym Phys 45:2912

Peng HS, Nie L, Chen DY, Jiang M (2007) Chem Commun p 2360

M, Minamikawa H (2005) Chem Rev 105:1401

isenberg A (1996) Langmuir 12:5980

(1998) Macromolecules 31:3509

bert A, Meier W (2004) Chem Commun p 1462

Qiu XP, Li Z (2002) Macromolecules 35:7742

(2000) Angew Chem Int Ed Engl 39:340

J, Liu FT, Tang BZ, Peng H, Pakhomov AB, Wong CY (2001) Angew 40:3593

21. Yan XH, Liu FT, Li Z, Liu GJ (2001) Macromolecules 34:9112
22. Hamley IW (2005) Block Copolymers in Solution: Fundamentals and Applications. Wiley, Chichester, West Sussex
23. Tuzar KZ, Kratochvil P (1993) Surf Colloid Sci 15:1
24. Zhang LF, Eisenberg A (1995) Science 268:1728
25. Cameron NS, Corbierre MK, Eisenberg A (1999) Canadian J Chem 77:1311
26. Ding JF, Liu GJ, Yang ML (1997) Polymer 38:5497
27. Bang J, Jain SM, Li ZB, Lodge TP, Pedersen JS, Kesselman E, Talmon Y (2006) Macromolecules 39:1199
28. Jain S, Bates FS (2004) Macromolecules 37:1511
29. Pochan DJ, Chen ZY, Cui HG, Hales K, Qi K, Wooley KL (2004) Science 306:94
30. Rodriguez-Hernandez J, Lecommandoux S (2005) J Am Chem Soc 127:2026
31. Canham PA, Lally TP, Price C, Stubbersfield RB (1980) Chem J Soc Faraday Trans 76:1857
32. Tao J, Stewart S, Liu GJ, Yang ML (1997) Macromolecules 30:2738
33. Won YY, Davis HT, Bates FS (1999) Science 283:960
34. Gohy JF, Lohmeijer BGG, Alexeev A, Wang XS, Manners I, Winnik MA, Schubert US (2004) Chem Eur J 10:4315
35. Discher DE, Eisenberg A (2002) Science 297:967
36. Ding JF, Liu GJ (1997) Macromolecules 30:655
37. Nagarajan R, Ganesh K (1989) J Chem Phys 90:5843
38. Dalhaimer P, Bates FS, Discher DE (2003) Macromolecules 36:6873
39. Wang XS, Guerin G, Wang H, Wang YS, Manners I, Winnik MA (2007) Science 317:644
40. Israelachvili J (1992) Intermolecular & Surface Forces. Academic Press, London
41. Napper DH (1983) Polymeric Stabilization of Colloidal Dispersions. Academic Press, London
42. Raez J, Manners I, Winnik MA (2002) J Am Chem Soc 124:10381
43. Raez J, Manners I, Winnik MA (2002) Langmuir 18:7229
44. Wang XS, Winnik MA, Manners I (2004) Angew Chem Int Ed 43:3703
45. Wang XS, Wang H, Frankowski DJ, Lam PG, Welch PM, Winnik MA, Hartmann J, Manners I, Spontak RJ (2007) Adv Mater 19:2279
46. Vilgis T, Halperin A (1991) Macromolecules 24:2090
47. Jenekhe SA, Chen XL (1998) Science 279:1903
48. Halperin A (1990) Macromolecules 23:2724
49. Wang XS, Wang H, Coombs N, Winnik MA, Manners I (2005) J Am Chem Soc 127:8924
50. Breiner U, Krappe U, Abetz V, Stadler R (1997) Macromol Chem Phys 198:1051
51. Bates FS, Fredrickson GH (1999) Phys Today 52:32
52. Yan XH, Liu GJ, Haeussler M, Tang BZ (2005) Chem Mater 17:6053
53. Li Z, Liu GJ (2003) Langmuir 19:10480

54. Fustin CA, Abetz V, Gohy JF (2005) Eur Phys J E 16:291
55. Gohy JF, Willet N, Varshney S, Zhang JX, Jerome R (2001) Angew Chem Int Ed 40:3214
56. Hu JW, Liu GJ (2005) Macromolecules 38:8058
57. Zheng RH, Liu GJ, Yan XH (2005) Am J Chem Soc 127:15358
58. Zheng RH, Liu GJ (2007) Macromolecules 40:5116
59. Mansky P, Liu Y, Huang E, Russell TP, Hawker C (1997) Science 275:1458
60. Kim SH, Misner MJ, Xu T, Kimura M, Russell TP (2004) Adv Mater 16:226
61. Liu GJ, Yan XH, Duncan S (2003) Macromolecules 36:2049
62. Liu GJ, Yan XH, Duncan S (2002) Macromolecules 35:9788
63. Liu GJ, Li Z, Yan XH (2003) Polymer 44:7721
64. Yan X, Liu G, Li H (2004) Langmuir 20:4677
65. Zimm BH, Crothers DM (1962) Proc Natl Acad Sci USA 48:905
66. Moore WR (1969) Prog Polym Sci 1:3
67. Yamakawa H, Fujii M (1974) Macromolecules 7:128
68. Yamakawa H, Yoshizaki T (1980) Macromolecules 13:633
69. Bohdanecky M (1983) Macromolecules 16:1483
70. Geng Y, Ahmed F, Bhasin N, Discher DE (2005) J Phys Chem B 109:3772
71. Won YY, Davis HT, Bates FS, Agamalian M, Wignall GD (2000) J Phys Chem B 104:9054
72. Won YY, Paso K, Davis HT, Bates FS (2001) J Phys Chem B 105:8302
73. Onsager L (1949) Ann NY Acad Sci 51:627
74. Flory PJ (1956) Proc R Soc London Ser A 234:73
75. Liu GJ, Ding JF, Qiao LJ, Guo A, Dymov BP, Gleeson JT, Hashimoto T, Saijo K (1999) Chem Eur J 5:2740
76. Werbowyj RS, Gray DG (1980) Macromolecules 13:69
77. Werbowyj RS, Gray DG (1976) Mol Cryst Liquid Cryst 34:97
78. Hunter RJ (1989) Foundations of Colloid Science, Vol. 1. Oxford University Press, Oxford
79. Milner ST (1991) Science 251:905
80. Ding JF, Tao J, Guo A, Stewart S, Hu NX, Birss VI, Liu GJ (1996) Macromolecules 29:5398
81. Tao J, Guo A, Stewart S, Birss VI, Liu GJ (1998) Macromolecules 31:172
82. Ziolo RF, Giannelis EP, Weinstein BA, Ohoro MP, Ganguly BN, Mehrotra V, Russell MW, Huffman DR (1992) Science 257:219
83. Yan XH, Liu GJ, Haeussler M, Tang BZ (2005) Chem Mater 17:6053
84. Prime KL, Whitesides GM (1991) Science 252:1164
85. Zheng RH, Wang JD, Liu GJ, Jao TC (2007) Macromolecules 40:7601
86. Ohno K, Koh K, Tsujii Y, Fukuda T (2003) Angew Chem Int Ed 42:2751
87. Liu GJ, Yan XH, Lu ZH, Curda SA, Lal J (2005) Chem Mater 17:4985
88. Stupp SI (2005) MRS Bull 30:546
89. Jin MH, Feng XJ, Feng L, Sun TL, Zhai J, Li TJ, Jiang L (2005) Adv Mater 17:1977
90. Vogelaar L, Lammertink R GH, Wessling M (2006) Langmuir 22:3125
91. Gao LC, McCarthy TJ (2006) Langmuir 22:2966
92. Koh K, Liu GJ, Willson CG (2006) J Am Chem Soc 128:15921
93. Liu G, Yan X, Li Z, Zhou J, Duncan S (2003) J Am Chem Soc 125:14039
94. Wang HD, Bash R, Yodh JG, Hager GL, Lohr D, Lindsay SM (2002) Biophys J 83:3619
95. Ozin GA, Arsenault AC (2005) Nanochemistry: A Chemical Approach to Nanomaterials. RSC Publishing, Cambridge
96. Yan X, Liu G, Li Z (2004) J Am Chem Soc 126:10059

97. Johnson DL, Esmen NA, Carlson KD, Pearce TA, Thomas BN (2000) J Aerosol Sci
    31:181
98. Price RR, Patchan M (1993) J Microencapsul 10:215
99. Price R, Patchan M (1991) J Microencapsul 8:301
100. Schnur JM (1993) Science 262:1131
101. Kim Y, Dalhaimer P, Christian DA, Discher DE (2005) Nanotechnology 16:S484
102. Silva GA, Czeisler C, Niece KL, Beniash E, Harrington DA, Kessler JA, Stupp SI (2004)
     Science 303:1352
103. Massey JA, Winnik MA, Manners I, Chan V ZH, Ostermann JM, Enchelmaier R,
     Spatz JP, Moller M (2001) J Am Chem Soc 123:3147
104. Garcia C BW, Zhang YM, Mahajan S, Di Salvo F, Wiesner U (2003) J Am Chem Soc
     125:13310
105. Kowalewski T, Tsarevsky NV, Matyjaszewski K (2002) J Am Chem Soc 124:10632
106. Tang CB, Tracz A, Kruk M, Zhang R, Smilgies DM, Matyjaszewski K, Kowalewski T
     (2005) J Am Chem Soc 127:6918

Adv Polym Sci (2008) 220: 65–121
DOI 10.1007/12_2008_145
© Springer-Verlag Berlin Heidelberg
Published online: 28 June 2008

Self-Assembled Polysaccharide Nanotubes Generated from β-1,3-Glucan Polysaccharides

Munenori Numata[1] (✉) · Seiji Shinkai[2]

[1]Department of Bioscience and Biotechnology, Faculty of Science and Engineering,
Ritsumeikan University, Kusatsu, Shiga 525-8577, Japan
numata@fc.ritsumei.ac.jp

[2]Department of Chemistry and Biochemistry, Graduate School of Engineering,
Kyushu University, Fukuoka 819-0395, Japan

Abstract β-1,3-Glucans act as unique natural nanotubes, the features of which are greatly different from other natural or synthetic helical polymers. The origin mostly stems from their strong helix-forming nature and reversible interconversion between single-strand random coil and triple-strand helix. During this interconversion process, they can accept functional polymers, molecular assemblies and nanoparticles in an induced-fit manner to create water-soluble one-dimensional nanocomposites, where individual conjugated polymers or molecular assemblies can be incorporated into the one-dimensional hollow constructed by the helical superstructure of β-1,3-glucans. The advantageous point of the β-1,3-glucan hosting system is that the selective modification of β-1,3-glucans leads to the creation of various functional one-dimensional nanocomposites in a supramolecular manner, being applicable toward fundamental nanomaterials such as sensors or circuits. Furthermore, the composites with functional surfaces can act as one-dimensional building blocks toward further hierarchical self-assemblies, leading to the creation of two- or three-dimensional nanoarchitectures.

Keywords Helical structure · Inclusion complex · Nanomaterials · Polysaccharide · Self-assemble

Abbreviations

AFM	Atomic force microscopy
ANS	1-Anilinonaphthalene-8-sulfonate
CD	Circular dichroism
CD	Cyclodextrins
CLSM	Confocal laser scanning microscopy
CMA	Carboxymethylated amylose
CUR	Curdlan
DPB	1,4-Diphenylbutadiyne
DMSO	Dimethyl sulfoxide
EDS	Energy dispersive X-ray spectroscopy
FITC	Fluorescein isothiocyanate
ICD	Induced circular dichroism
MA	2,3-O-Methylated amylose
ORD	Optical rotatory dispersion
PANI	Poly(aniline)
PEDOT	Poly(3,4-ethylenedioxythiophene)
PMDS	Permethyldecasilane
Poly(C)	Polycytidylic acid
PPE	Poly(p-phenylene ethynylene)
PT	Poly(thiophene)
PS	Poly(silane)
SEM	Scanning electron microscopy
SPG	Schizophyllan
SWNT	Single-walled carbon nanotubes
TEM	Transmission electron microscopy
TMPS	Trimethoxypropylsilane
TNS	2-p-Toluidylnaphthalene-6-sulfonate
VIS-NIR	Visible-near-infrared

1
Introduction

1.1
Polysaccharide Nanotubes

Polysaccharides are abundant organic compounds in nature, constituting a significant portion of our diet, and serve mainly as energy stores, e.g., starch in plants and glycogen in animals. With regard to potential pharmaceutical effects of polysaccharides, they capture intense attention as "biomaterials". Besides, the inherent bio-degradable nature of polysaccharides makes them potential candidates for "eco-materials", such as green-plastics. The peculiar physical and chemical properties of polysaccharides certainly play a significant role in animals and plant cells. For example, some polysaccharides possess inherently stiff and rigid structures, which cause mechanical strength, like cellulose in plants, peptidoglycans in bacteria and chitin in the exoskeletons of arthropods. Furthermore, specific saccharide chains serve as cell surface recognition signals for antibodies, hormones, toxins, acting as indispensable biopolymers for our life through strict molecular recognition.

Nature produces numerous kinds of polysaccharides in an appropriate biological environment. The structural diversity of the natural polysaccharide is fully commensurate with a diverse array of molecules that can be generated from only a limited number of monosaccharides as building blocks by linking them in a variety of ways. This structural feature of polysaccharide characterized by diversity is in sharp contrast to that of other natural polymers such as polynucleotides and peptides with very regular, uniform and well-identified nanostructures.

Recently, the greatest growth has been achieved in the structured determination of natural polysaccharides, and hence this has increased our understanding in relation to structural features as well as functionalities of polysaccharides [1]. The basic knowledge of the structural feature of polysaccharides is essential toward the application as fundamental nanomaterials. For example, X-ray diffraction patterns of various natural polysaccharides have revealed that some of them adopt well-defined helical nanoarchitectures such as polynucleotides, which have never been produced through an artificial polymerization reaction, encouraging us to pursue the possibilities of natural nanotubes.

1.1.1
Amylose

Amylose is a most familiar polysaccharide with regular helical structures defined at the nanoscale (Fig. 1). Carbohydrates are stored in the starch form,

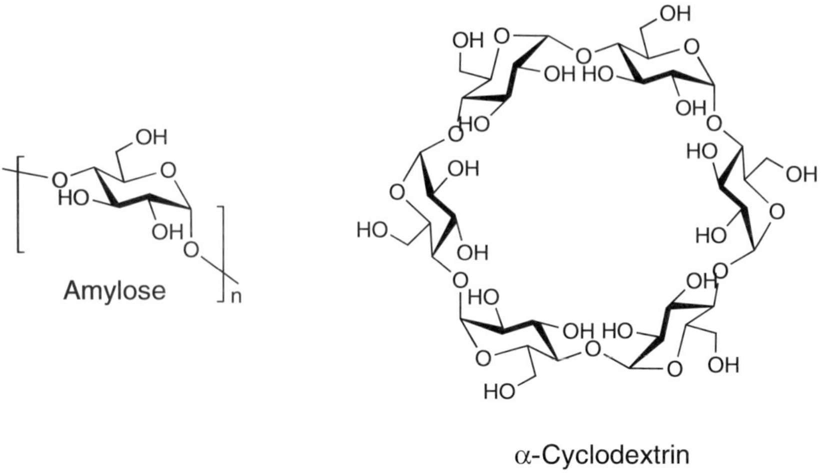

Fig. 1 Repeating units of amylose and α-cyclodextrin

which is a mixture of amylose and amylopectin. Amylopectin is a branched polymer, whereas amylose is essentially a linear polymer of (1 → 4) linked α-D-glucose units, i.e., the unbranched portion of starch. Natural amylose exists as a double-stranded parallel helix, involving two main polymorphs, whereas a single-stranded amylose can be seen in DMSO (Fig. 2). These amylose polymorphs commonly take the same helical parameters, i.e., diameter of 1.3 nm and pitch of 0.8 nm involving six glucose residues per turn [2–4]. This structural feature of amylose can be regarded as one of the typical helical nanotubes, and actually amylose forms the well-known "blue complex" with iodine [5, 6]. In addition, various low molecular organic molecules, such as DMSO, n-butanol and n-pentanol are also entrapped in the amylose helical cavity [7–10]. The main driving force for the guest binding is hydrophobic interaction that occurs only in the hydrophobic amylose cavity. On the basis of these fundamental investigations on the potential hosting ability of amylose, so far, several research groups have independently demonstrated that amylose can entrap various functional molecules or polymers, leading to the creation of one-dimensional supramolecular complexes. For example, Kim et al. have reported that hydrophobic cyanine dyes with alkyl tails are entrapped into the helical cavity to give water-soluble complexes [11, 12]. The hydrophobic nature in the dye molecules with long alkyl tails is indispensable for the formation of the stable inclusion complexes. The inclusion of the highly polarized dye molecules is of interest in the light of materials science. For example, the supramolecular dye assemblies created with the assistance of the amylose host show unusual thermochromic or optical behaviors, such as second-harmonic generation effects. Once guest dye molecules are entrapped into the cavity, stretching out along the helix axis, the conformational

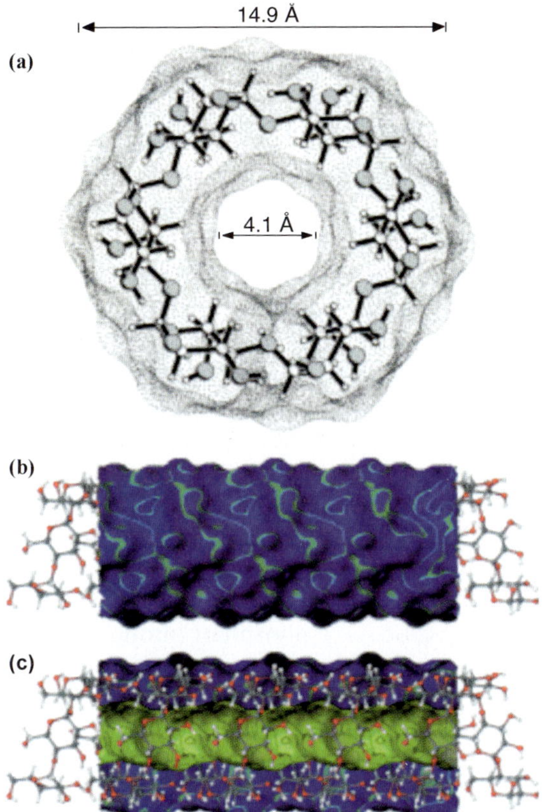

Fig. 2 Example of crystallographic conformation of V_H-amylose. (**a**) Hydrophilic outer surface and (**b**) Inner hydrophobic cavity in *yellow* (**c**). Reprinted with permission from [5]

freedom of dye molecules drastically decreases owing to host–guest interactions, resulting in enhanced fluorescence intensity or unusual thermochromic behavior.

Oligosilanes are σ-conjugated conducting polymers, possessing unique electronic properties due to their electron delocalization along the main chain. Sanji and Tanaka reported that partially carboxymethylated amylose (CMA) can entrap the oligosilane in the helical cavity to give insulated wire-like architectures [13, 14]. The chemical treatment of natural amylose weakens the intra- or interpolymer hydrogen-bonding interactions, so that amylose assumes a somewhat flexible conformation, accompanied with improvement of water-solubility. The powder X-ray diffraction studies revealed that CMA adopts an 8_1-helical structure after incorporation of oligosilane, indicating that the cavity of CMA is somewhat enlarged by the guest inclusion. From CD spectroscopic studies of the complex, the incorporated oligosilane adopts the helical conformation influenced by the chiral cavity of CMA, forming a diastereomeric complex. Similar phenomena are also observed for oligothiophene/CMA systems, that is, oligothiophenes are insulated into the chiral CMA cavity which forces the oligothiophene to adopt a helical conformation [15].

Akashi et al. have demonstrated that chemically modified amylose, i.e., partially 2,3-*O*-methylated amylose (MA), also exhibits excellent hosting ability toward polymer guests such as poly(tetrahydrofuran) and poly(ε-caprolacton) [16]. The helix content of MAs, which is regarded as a measure of the helix-forming ability, could be tuned by the methylation ratio. They investigated the potential abilities to form nanotubes for a series of MAs with different methylation ratios. Consequently, MAs with a moderate methylation ratio tend to form stable inclusion complexes with the selected guest polymers, suggesting the view that these MAs contain some continuous helical segments that play an essential role for improving the inclusion ability of amylose toward guest polymers.

An alternative strategy toward the formation of such inclusion complexes, reported by Kadokawa et al., is to use the enzymatic polymerization of glucose on a template polymer, where phosphorylase catalyzes a polymerization reaction of α-D-glucose 1-phosphate monomer along the template polymer [poly(tetrahydrofuran)] in a twisting manner [17]. The same inclusion complex was not formed upon just mixing natural amylose with polymer [poly(tetrahydrofuran)], probably due to the rigid conformation of natural amylose.

In view of the creation of functional inclusion complexes, single-walled carbon nanotubes (SWNTs) would be an ultimate target polymer due to their potential as fundamental nanomaterials. Aiming at the practical application of SWNTs as functional polymers, the difficult handling arising from their poor solubility and strong cohesive nature is a serious problem for chemists. Accordingly, formation of the inclusion complex with SWNTs would be a new path to accelerate development of carbon nanotube chemistry. The first example is SWNT/amylose composite formation reported independently by Kim et al. and by Stoddard et al. [18, 19]. It was shown therein that SWNTs can be incorporated into the amylose cavity to form stable inclusion complexes. Considering the diameter of SWNTs (1–2 nm), the cavity size of natural amylose (ca. 0.4 nm) would be changed by the inclusion of SWNTs.

Natural amylose and its chemically modified form, i.e., CMAs and MAs, possess basically a rigid cavity, and therefore, they can include guest molecules or polymers in a size-selective manner because the cavity can act flexibly in an induced-fit manner, being different from those of cyclodextrines (CDs). From the stand point of the nanotube structure, this is in contrast to CD hosts in rigidity and ring structures.

1.1.2
Cyclodextrin Arrays

Cyclodextrins (CDs) are cyclic compounds consisting of six to eight glucose units; i.e., α-, β- and γ-CDs (Fig. 1). The cavity sizes of these CDs are 0.43 nm, 0.60 nm, and 0.74 nm, respectively [20]. These cyclic cavities are constructed with $\alpha(1 \rightarrow 4)$-linked glucose units in the same fashion as for the amylose cavity. Once the polymeric guest threads through the CD cavities, CDs are assembled into array structures, which are now called pseudo-polyrotaxane structures. One may regard them as a sort of self-assembled carbohydrate nanotube [21–23]. Therefore, we herein overview CD arrays as a potential self-assembled nanotube, characterizing the difference between CDs and β-1,3-glucan hosting systems.

CDs are known to form inclusion complexes with various low-molecular weight compounds, including nonpolar aliphatic molecules, polar amines and acids as well as various polymers [24]. Along this line, CDs also have ideal structures for the construction of molecular nanotubes [25–28]. Harada et al. have successfully demonstrated that a series of CDs are threaded onto a polymer chain to give pseudo-polyrotaxane structures [29–34]. For example, α-CDs form a polyrotaxane structure with poly(ethylene glycol), which is obtained as a crystalline precipitate. The powder X-ray diffraction pattern of the precipitate led to the conclusion that α-CDs are tightly packed on the polymer in a head-to-head manner. It is worth mentioning that β-CDs and γ-CDs also exhibit similar hosting abilities as α-CDs, the size correlation between their cavity size and the diameter of guest polymers being well recognized: β-CDs form complexes not with poly(ethylene glycol) but with poly(propylene glycol), which is slightly larger than poly(ethylene glycol) in its diameter. Similarly, γ-CDs form complexes with poly(methyl vinyl ether), whereas α-CDs and β-CDs do not give complexes with this polymer. Strictly speaking, however, these polyrotaxanes may not be "true" nanotubes because CD units are not connected by covalent bonds. Interestingly, the cross-linking of the adjacent CDs on a template polymer leads to the creation of complete tubular structures even after removing the template, where temporarily formed tubular structures are fixed on the template polymer through the chemical treatment.

After the initial research on CD nanotubes by Harada et al. and by Wenz et al. the intriguing systems have been widely extended to various functional polymers such as conjugated polymers (Fig. 3d), including poly(aniline) (PANI), poly(thiophene) (PT), poly(silane) (PS), etc., which can also thread through the CD cavities to form the polyrotaxane-type complexes (Fig. 3b) [35–50]. In view of the fundamental research on the conjugated polymers, the advantageous point of the CD nanotube systems is that the functional polymers can be well insulated as an individual polymer chain even in their solution and in the solid state. Since the conjugated polymers

(a)

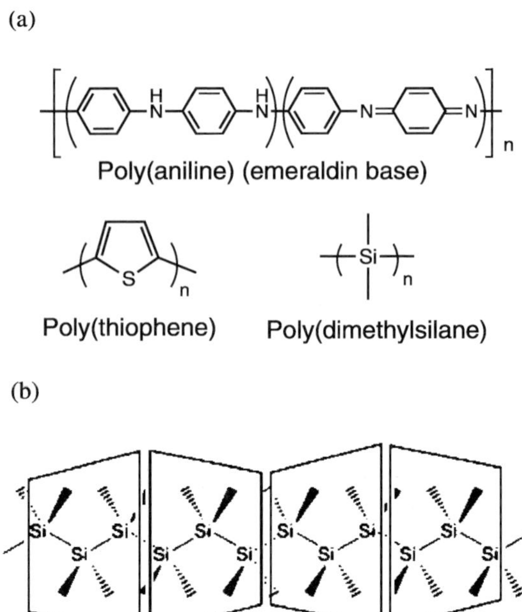

(b)

Fig. 3 Example of π- and σ-conjugated polymer guests threading into the CD cavity (*top*) and example of self-assembled γ-CD nanotube including poly(dimethylsilane). Reprinted with permission from [49]

have a strong tendency to form an insoluble aggregate mass mainly due to their strong stacking nature, the insulated polymer would provide dramatic advantages in relation to the stability as well as the electronic and photochemical properties of the guest polymers.

1.2
β-1,3-Glucans

1.2.1
Structural Aspects of β-1,3-Glucans

β-1,3-Glucans are present in a number of fungi, where they function as a structural polysaccharide like cellulose, e.g., extracellular microbial polysaccharide and they are essentially a linear polymer of $(1 \rightarrow 3)$-linked β-D-glucose units (Fig. 4) [51–65]. Among a series of β-1,3-glucans, curdlan (CUR) is known as one of the simplest β-1,3-glucans [60–65]. X-ray diffraction patterns of CUR in the anhydrous form revealed that it adopts a right-handed 6_1 triple helix with a diameter of 2.3 nm and pitch of 1.8 nm (Fig. 5) [61, 65]. In contrast to the simple CUR structure, SPG has side glucose groups linked at every third main-chain glucose. The side glucose groups

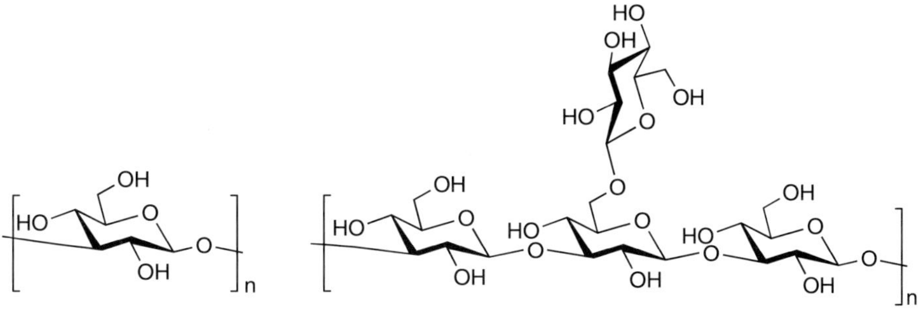

Curdlan (CUR)

Schizophyllan (SPG)

Fig. 4 Repeating units of β-1,3-glucans: CUR and SPG

endow SPG with water solubility, whereas they do not affect the helical conformation of the main chain. The helical parameters of β-1,3-glucans are almost consistent with those of double-stranded DNA, but somewhat larger than those of amylose, indicating that CUR adopts a wide helical structure. Atkins et al. have revealed by X-ray diffraction patterns that in the triple helix structure of CUR, three glucoses composed of different chains are bound together through the hydrogen bonds among the three O-2 atoms, forming the one-dimensional hydrogen-bonding network in the triple helix [64]. This structural analysis revealed that in the triple helix there is not enough space to accommodate even small molecular guests. Therefore, the sole strategy for β-1,3-glucans to exert their hosting ability is to dissociate the original triple helix into a single strand, losing the inner hydrogen-bonding network: these dissociated β-1,3-glucan chains would possess potential ability to include guests and to form nanotubes.

Norisuye et al. have reported the interesting solution properties of β-1,3-glucans where denaturing and renaturing of the triple helix can take place reversibly; that is, schizophyllan (SPG) dissolve in water as a triple helix (t-SPG), whereas as a single chain in DMSO (s-SPG). When water is added to the DMSO solution, however, renaturing of s-SPG is promoted, resulting in the formation of the original triple helix (Fig. 5d) [51, 59, 61, 66, 67]. CUR with an appropriate molecular weight also exhibits a similar reconstructing ability of the triple helix.

1.2.2
Complexation Between β-1,3-Glucans and Polynucleotide

Recently, Sakurai et al. reported the first example of β-1,3-glucan forming a macromolecular complex with some polynucleotides; that is, when the renaturing of β-1,3-glucan was carried out in the presence of some polynucleotides, it formed novel triple-stranded structures consisting of two β-1,3-

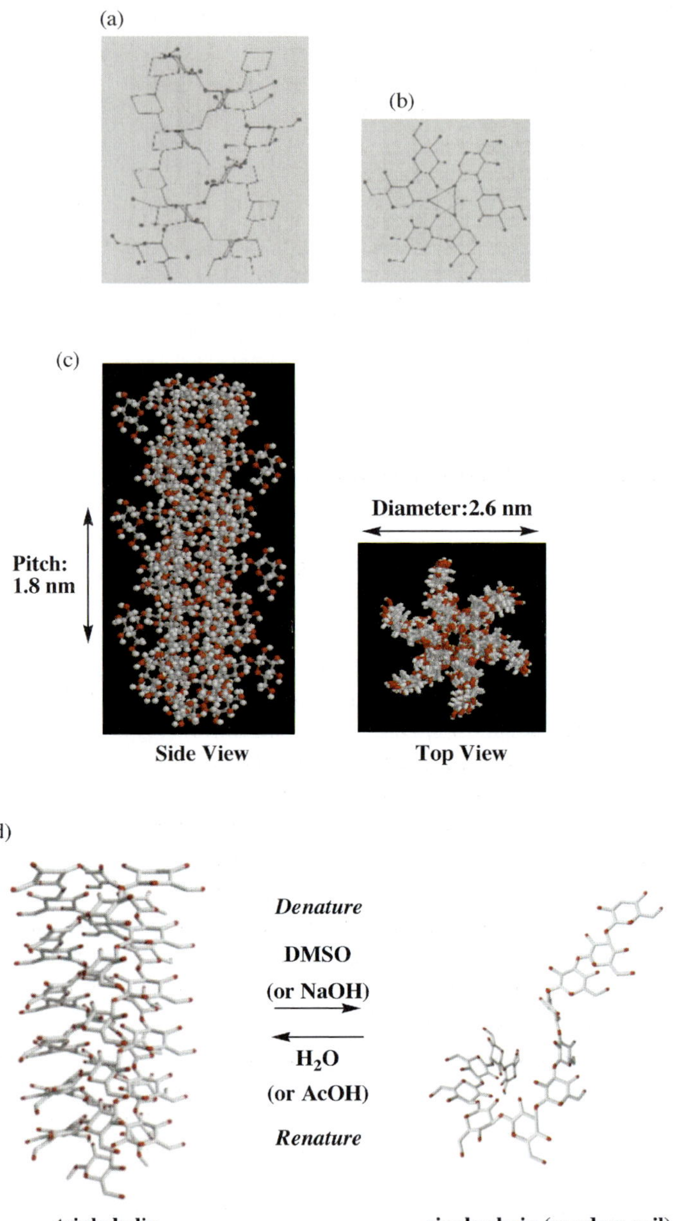

Fig. 5 Projection of the CUR triple helix **a** in the x–z plane and **b** in the x–y plane, showing the hydrogen-bonding network constructed in the helix (hydrogen atoms are omitted for both structures) (Reprinted with permission from [64]). **c** Calculated SPG triple helix structures based on the crystal structure of CUR and **d** Schematic illustration of denature/renature processes

glucan chains and one polynucleotide chain [68, 69]. These findings imply that β-1,3-glucan can interact with guest polymers only in its dissociated state, which would allow the glucose main-chain to adopt various conformations different from the original triple helix. The main driving force for the formation of the macromolecular complexes is considered to be hydrophobic interaction in addition to hydrogen-bonding interactions occurring inside the novel triple helix. These intriguing findings encouraged us to apply this unique and unusual hosting system to other functional polymeric guests, where the denaturing and renaturing processes play a significant role in inclusion of various functional materials, such as polymers, molecular assemblies, inorganic particles, etc.

2
Unique One-Dimensional Hosting Abilities

2.1
Inclusion of Single-Walled Carbon Nanotube (SWNT)

Since the discovery of single-walled carbon nanotubes (SWNTs) by Iijima, they have been regarded as ideal nanomaterials due to their unique electronic, photochemical, and mechanical properties [70–72]. Much effort has been paid to apply SWNTs as practical nanomaterials, however the strong cohesive nature and poor solubility of SWNTs have caused a serious problem; that is, these properties hamper work on SWNTs as "functional polymers". As a potential solution to this problem, it would be worth wrapping SWNTs with synthetic or biological polymers, promoting dissociation of the SWNT bundle to give a homogeneous solution without damaging SWNT surfaces [73–79]. In particular, a natural polysaccharide such as amylose is an excellent solubilizer for SWNTs, because polysaccharides have no light absorption in the UV-VIS wavelength region, which is suitable for exploiting the photochemical properties of the resultant composite.

The main driving forces for the reconstruction of the triple-stranded β-1,3-glucans from the single-stranded ones are considered to be the hydrophobic interaction in addition to the hydrogen-bonding interactions. On the basis of this experimental fact, it is expected that SWNTs might be entrapped in the inside hollow of the β-1,3-glucans helical structure mainly owing to the hydrophobic interaction (Fig. 6). Unlike the amylose hosting system, however, natural β-1,3-glucans do not have enough cavity to accommodate SWNT with 1–2 nm diameters. Thus, some conformational change in the β-1,3-glucan main-chain would be needed after entrapping the guest polymers.

As a preliminary experiment to investigate whether β-1,3-glucans can actually entrap such a rigid polymer into their cavities, SWNTs were cut into an appropriate length (1–2 μm) by acid treatment, which makes handling of

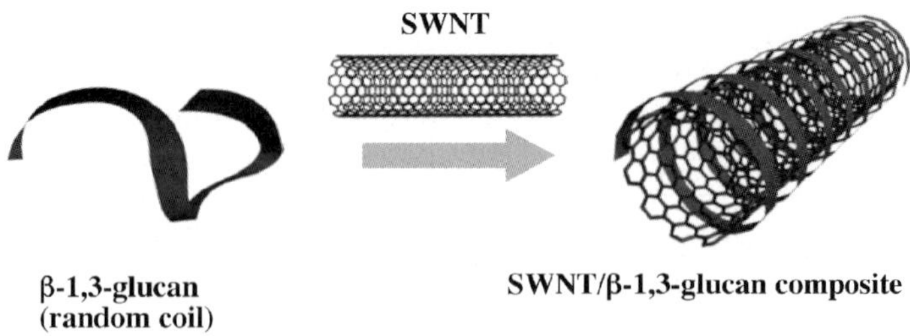

Fig. 6 Schematic illustration of the formation of SWNTs/β-1,3-glucan composite

SWNTs easy [80–83]. As a result, the cut SWNTs (c-SWNTs) can be easily dispersed into water, but they still tend to form bundle structures several tens of nanometers in diameter. Consequently, an s-SPG (M_w = 150 kDa) DMSO solution was directly added to an aqueous solution containing the bundle c-SWNTs, according to the same procedure as for the polynucleotide guests, expecting that SWNTs are entrapped into the SPG cavity [68]. To remove an excess amount of SPG feed, the resultant solution was subjected to centrifugations. The presence of c-SWNT in the obtained aqueous solution was evidenced by the measurements of VIS-NIR and Raman spectroscopy. The direct evidence that SWNTs are really entrapped into the SPG cavity was obtained by Atomic Force Microscopy (AFM). Interesting, when c-SWNTs were dispersed with the aid of SPG into water, the surface of the fibrils showed a periodical structure with inclined stripes, reflecting the strong helix-forming nature of the β-1,3-glucan main-chain (Figs. 7 and 8). Subsequently, the periodical interval in the helical stripes was estimated by AFM. It was confirmed from scanning along a fibril that the mountain and the valley appear periodically at 16 nm intervals. In addition, from the height profile analysis by AFM, it was revealed that most composites are ca. 10 nm in height, indicating that β-1,3-glucans can wrap around bundle c-SWNTs, with a change in their original helix parameters. As a reference experiment, on the other

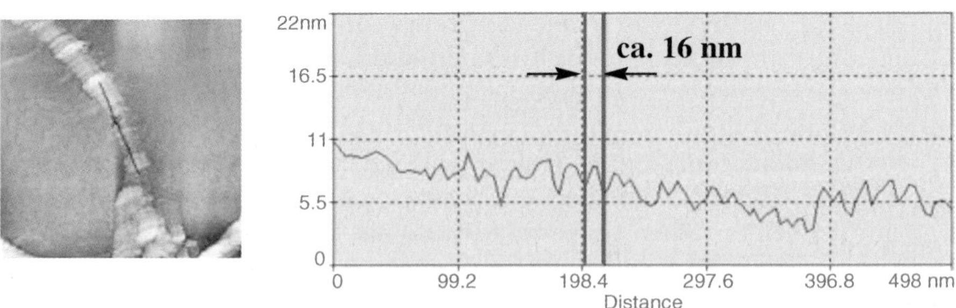

Fig. 7 AFM image of c-SWNTs bundle/SPG composite and its scanning profiles of vertical section plans. Reprinted with permission from [86]

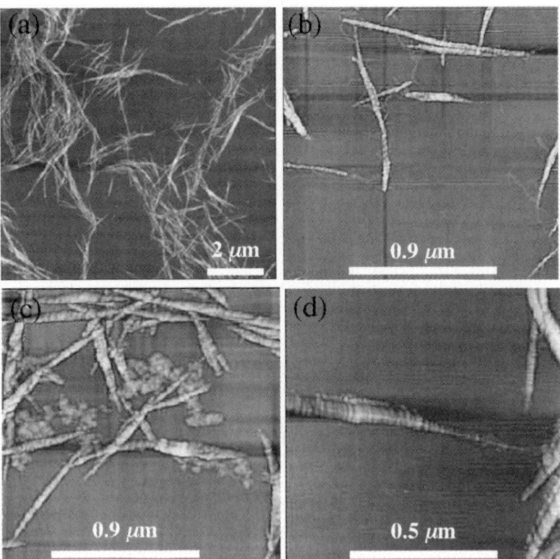

Fig. 8 AFM images of **a** c-SWNTs, **b** c-SWNTs/SPG composite, **c** c-SWNTs/CUR composite, **d** Magnified picture of fibrils in s-SWNTs/SPG composite. Amorphous structure observed around the composite is considered to be uncomplexed CUR which cannot be removed during centrifugation process due to its poor solubility in water

hand, when a DMSO solution of s-SPG was cast on mica, we could observe a fine polymeric network structure. Besides, the surface of c-SWNTs themselves did not give any specific patterns as seen in the composite surface. These results clearly show that β-1-3-glucan can act as a one-dimensional host for SWNTs, where the characteristic helical structure of β-1-3-glucans is reconstructed on the SWNTs surface [84]. The β-1,3-glucan one-dimensional host is characterized by this well-identified wrapping mode arising from the strong helix-forming nature of β-1,3-glucans; that is, the hydrophobic inner surface of β-1,3-glucans interact with SWNTs, whereas a hydrophilic surface exists on the composite surface. It should be noted that only the composites containing bundle SWNTs are intentionally highlighted for microscopic observation because the helical structure constructed on such a composite can be easily recognized by microscopic techniques.

Natural CUR is scarcely soluble in water due to the lack of side glucoses. Nevertheless, when a CUR chain was cut into a moderate length, e.g., several tens of thousand, by formic acid-hydrolyzed treatment [85], the resultant CUR acts as a one-dimensional host like SPG. Actually, when s-CUR was used as a wrapping agent instead of s-SPG, a similar periodical structure as seen in the c-SWNT/SPG composite can be observed on bundled SWNTs (Fig. 8c) [84]. The findings lead to the conclusion that the $\beta(1\rightarrow3)$ glucose linkages are indispensable for the unique hosting capabilities of β-1,3-glucans.

The novel SWNTs/β-1,3-glucan composites thus obtained were thoroughly characterized by several spectroscopic and microscopic methods including HRTEM, EDS-TEM, SEM, CLSM, which consistently support the view that the β-1,3-glucans really wrap SWNTs.

SWNTs exhibit their unique electronic and photonic properties only when they exist as an individual fiber. Taking the fact that SPG and CUR adopt the tight helical structure in their natural state, they are expected to wrap one-piece of SWNTs into their hydrophobic cavity, avoiding the formation of the unfavorable bundle structure. Here, if we intend to apply the resultant composite to practical nanomaterials, as-grown SWNTs (ag-SWNTs) are favorable because of no electrochemical defect on the SWNT surface. Although, unlike c-SWNTs, ag-SWNTs have a strong tendency to form the bundle structure, they are easily dispersed into water after entrapping into the β-1,3-glucan cavity with the aid of sonication. This remarkable solubilization capability even for ag-SWNTs allows us to investigate the detailed properties of the resultant composites [86]. First, the ag-SWNTs/SPG composite was subjected to VIS-NIR measurements in D_2O solvent (Fig. 9). The characteristic sharp bands assignable to individual SWNTs can be observed in the VIS-NIR region, supporting the view that one or a few pieces of ag-SWNTs are included in the SPG helical structure. AFM observations support the view more quantitatively. From the height profile of the composite, it can be recognized that most of them are 2–3 nm in diameter and the distribution is very narrow. Furthermore, when the surface of the composite was scanned along the fibril, a periodical pattern as seen in the c-SWNT/SPG

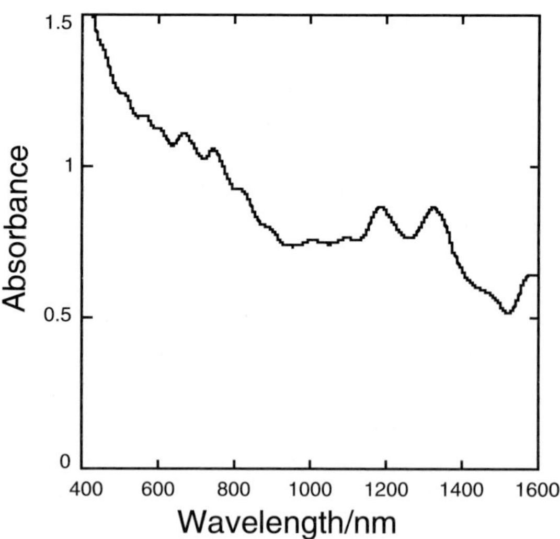

Fig. 9 VIS-NIR spectrum of ag-SWNTs/SPG solution: D_2O, cell length 1.0 cm , room temperature. Reprinted with permission from [86]

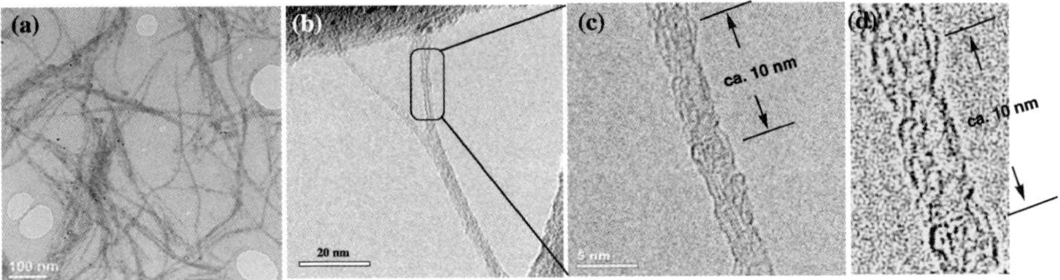

Fig. 10 **a** TEM image of ag-SWNT/SPG composite, and **b**, **c** its magnified picture. **d** The original image of (**c**) was Fourier filtered to enhance the contrast of the composite. Reprinted with permission from [86]

composite can be recognized. Subsequently, the ag-SWNTs/SPG composite was directly characterized by high-resolution TEM (HRTEM). From Fig. 10, the composites are visually recognized to be a very small, fibrous structure but not a bundled structure. Moreover, clearly recognized in the Fourier filtered image was that two s-SPG chains twine around one ag-SWNT, providing "decisive" evidence that one piece of ag-SWNTs is wrapped by s-SPG chains. The helical pitch is estimated to be ca. 10 nm, which is longer than that of the original triple helix, suggesting the view that the conformation of SPG chains is changeable with incorporation of guests, in sharp contrast to CDs or amylose systems. These conformational changes in the main-chain would be allowed only for the single- or double-stranded helix and therefore the denaturation process of the original triple helix should be indispensable for the guest inclusion. Recently, Coleman and Ferreira presented a simple and general model that describes the ordered assembly of polymer strands on nanotube surfaces, the energetically favorable coiling angles being estimated to be 48–70° [87]. This theoretical approach is partly complementary to our present results, but their single chain wrapping system cannot be directly compared with our two-chain wrapping system.

Today, electron tomography is expected to play a central role in characterization of composite nanomaterials, because it can provide a real three-dimensional distribution map of constituent materials [88–93]. This imaging technique was applied to our composite nanomaterials consisting of two nanofibers. The three-dimensional structures are reconstructed from an angular series of two-dimensional structure in Fig. 10b. The four selected stills are shown in Fig. 11, where ag-SWNT appears as a white fine fibril (for example, the right branch in the 45° rotating image) and the SPG thin layer wraps the fibril. The s-SPG chains exist as a uniform layer around one or two pieces of ag-SWNTs even viewing from any angle. This fact implies that s-SPG *wraps* one piece of ag-SWNT.

If β-1,3-glucans maintain their characteristic helix-forming nature even on the SWNT surface, the dissociation from SWNTs would be promoted by add-

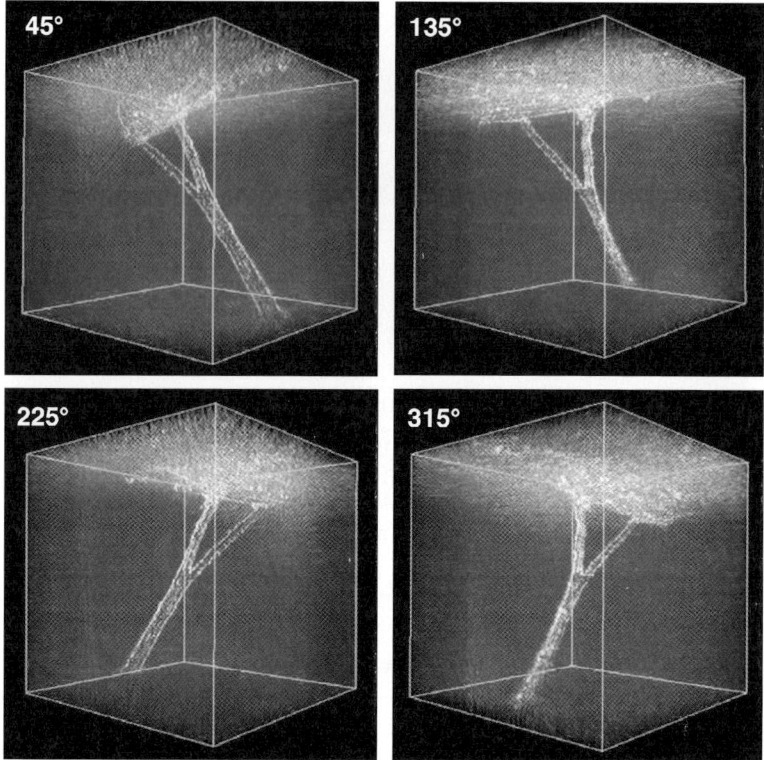

Fig. 11 Reconstructed three-dimensional images of ag-SWNTs/SPG composite, viewed from different orientations. These correspond to the clockwise rotation images of the two-dimensional image in Fig. 10b, rotating around a perpendicular axis with 45, 135, 225 and 315°, respectively. Reprinted with permission from [86]

ition of DMSO or NaOH, in which β-1,3-glucans exist as a single chain. As shown in Fig. 12, when DMSO was added to the aqueous solution containing the composite adjusting the final composition to be 50% v/v, the entrapped ag-SWNTs were immediately precipitated out. This result gives the strong impression that the composites are stabilized by noncovalent interactions occurring between the hydrophobic core and the hydrophilic shell, which can be easily peeled off by the various chemical stimuli.

As a summary of the forgoing findings, particularly interesting are the facts that (1) the periodical stripe structure, which stems from a helical wrapping mode characteristic of β-1,3-glucans, is confirmed, and (2) β-1,3-glucans have potential abilities to accommodate various guest polymers due to their flexible conformational changes, without being affected by the diameter as well as the chemical properties of guest polymers.

Fig. 12 Picture of aqueous ag-SWNTs/SPG composite solution (*left*) and that taken after addition of DMSO (*right*): The final solution contains 50% v/v DMSO. Addition of aqueous 1.0 M NaOH solution results in a similar precipitation of ag-SWNTs. Reprinted with permission from [86]

2.2
Inclusion of Conjugated Polymers

2.2.1
Inclusion of Poly(aniline) (PANI)

PANI is one of the most promising and widely studied conductive polymers owing to its high chemical stability, high conductivity, and unique redox properties [94–96]. In spite of these advantages, PANI and its derivatives are hardly applicable for conductive nanowires in a bottom-up manner, because these PANIs tend to form amorphous aggregates composed of highly entangled polymeric strands. Much research effort has therefore been put into the manipulation of an individual PANI nanofiber and in the fabrication of parallel-aligned strands to show excellent conductivity through the nanofibers.

From the forgoing c-SWNTs/SPG composite system, it would be expected that β-1,3-glucans can fabricate PANI nanofiber structures several tens of nanometer in diameter, acting as a one-dimensional host for PANIs bundles. The objective PANIs/SPG nanocomposites can be prepared through gradual dilution of a DMSO solution containing commercially available PANI (emerardine base, $M_w = 10$ kDa) and s-SPG ($M_w = 150$ kDa) with water followed by purification through repeated centrifugation [97]. TEM observations showed the formation of fibrous architectures from PANIs/SPG nanocomposites whose lengths were consistent with that of the used SPG (Fig. 13). The observed nanofibers are highly contrasted without any staining due to the adsorption of the electron beam. Importantly, the contrast,

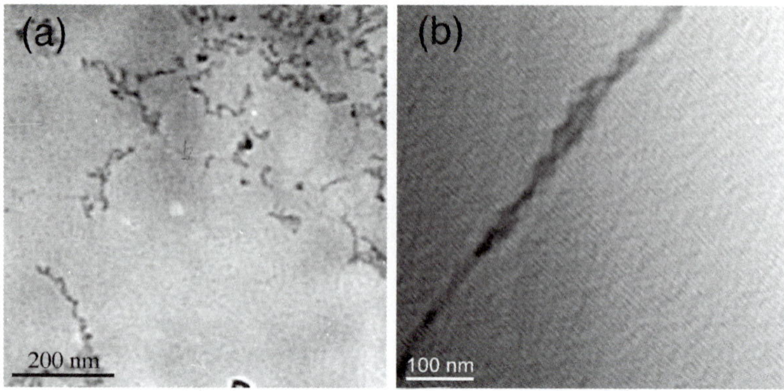

Fig. 13 a TEM images of PANIs/SPG composites, **b** Magnified image of (**a**), the images were taken without staining. PANIs themselves afforded spherical massive aggregates. Reprinted with permission from [97]

which should arise from incorporated PANIs, continuously exists all over the range of the obtained composite. Since the length of the used PANIs is rather shorter than that of s-SPG, the incorporated PANIs fibers are bundled into the one-dimensional fibers along the SPG cavity. The diameter of the smallest fiber is estimated to be around 10 nm, indicating that several PANI strands are co-entrapped within the one-dimensional cavity. Amylose is also considered to be a potential host for the creation of such an insulated wire. As far as our investigation, however, amylose could not form such a fine fi-

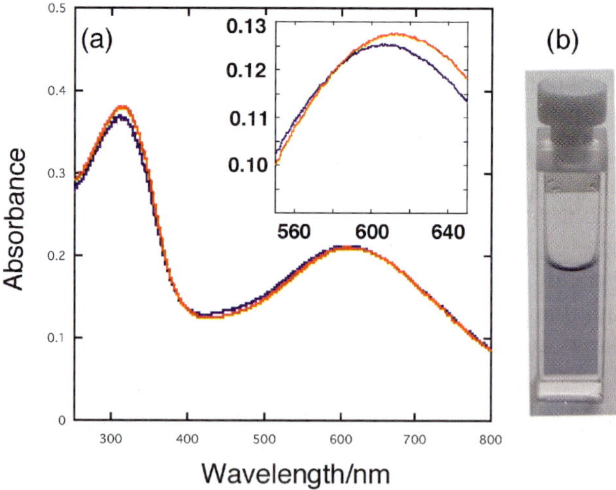

Fig. 14 a UV-VIS spectra of PANIs/SPG (*blue line*) and amylose/PANIs (*red line*) aqueous solution, room temperature, cell length 0.5 cm, **b** Photo image of PANIs/SPG aqueous solution. Reprinted with permission from [97]

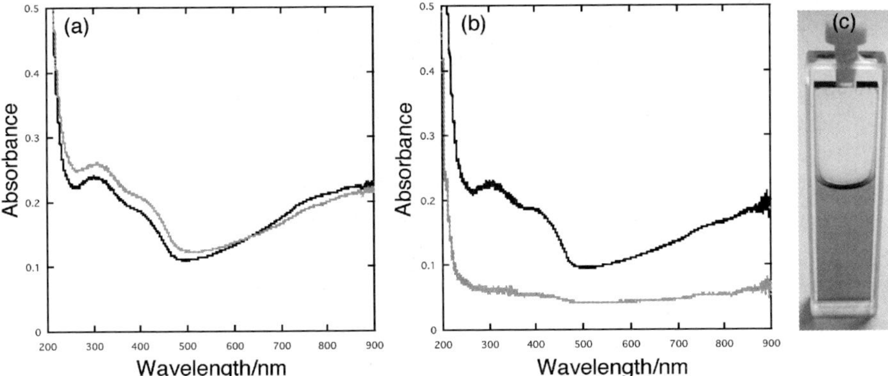

Fig. 15 UV-VIS spectra of **a** PANIs/SPG composite (*black line*) and PANIs/amylose composite (*gray line*) after H$^+$ doping, **b** After 24 h under the same condition and **c** Photo image of PANIs/SPG composite aqueous solution after doping. Reprinted with permission from [97]

brous structure upon mixing with PANIs according to the same procedure. The PANIs/amylose mixture did not give any significant specific structure in TEM.

Spectroscopic measurements of the composite support the view that PANIs bundles are entrapped into the one-dimensional cavity, which forces PANI fibers to arrange in a parallel fashion, leading to the creation of a nanowire structure (Fig. 14). Furthermore, the obtained PANIs/SPG composite was easily doped by the HCl treatment and the resultant solution was stable for several weeks (Fig. 15). On the other hand, the amylose/PANIs composite formed a precipitate after the same treatment. The morphology of the composite was scarcely changed through this doping process. Taking the fact into account that the diameter of the composite is rather larger than that of native t-SPG, SPG would wrap PANIs bundles in a somewhat disordered conformation, which would allow the dopant to reach the PANIs surface. This system would be useful for fabricating PANI nanofibers in a supramolecular manner.

2.2.2
Inclusion of Poly(thiophene) (PT)

So far, several intriguing approaches have been developed to fabricate an individual PT fiber, including a covalent and noncovalent approach in which the PT backbone is shielded by wrapping within a dendritic wedge or threading PT through CDs sheath [98–102]. Along this line, the composite formation between SPG and PT is of interest as a new approach leading to the creation of novel PT molecular wires [103]. To prepare a well-ordered one-dimensional nanocomposite, the strong π–π stacking interactions among PT backbones

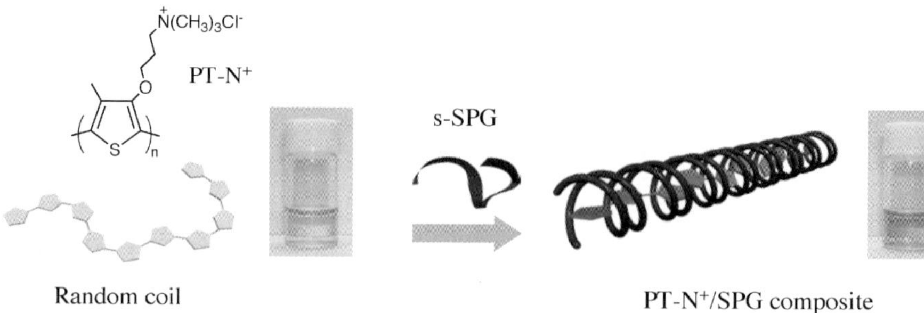

Fig. 16 Schematic illustration of the chirally insulated PT-N$^+$ in the helical SPG cavity

should be suppressed before being insulated into the β-1,3-glucan cavity. In the present study, water-soluble cationic polythiophene (PT-N$^+$) was synthesized for the well-identified composite formation, because it would interact with β-1,3-glucan as an individual polymer. Expectedly, the polycationic nature of PT-N$^+$ and the resultant weakened packing mode would result in a well-characterized PT-N$^+$/SPG complex, in which only one PT strand is encapsulated within the one-dimensional cavity of SPG (Fig. 16). Figure 17a and b shows absorption and emission spectra, comparing the PT-N$^+$ and PT-N$^+$/SPG composites. The absorption maximum of PT-N$^+$ appears at 403 nm, whereas that of PT-N$^+$/SPG is drastically red-shifted to 454 nm by ca. 50 nm, demonstrating that SPG forces the PT backbone to adopt a more planar conformation which increases the effective conjugation length. In the general case, when PTs form π-stacked aggregates in poor solvent, the UV-VIS spectrum is characterized by the appearance of a vibronic band in the longer wavelength region [104]. In the present system, however, such a peak attributable to the stacked aggregate was not observed.

The emission maximum (561 nm) of PT-N$^+$/SPG composites is also redshifted from that of free PT-N$^+$ (520 nm) and slightly increases in intensity, supporting the view that the PT-N$^+$ backbones become more planar and more isolated in the SPG cavity. It should be emphasized that no red shift of the absorption peaks is observed for the cast films of PT-N$^+$/SPG composite, indicating that the PT-N$^+$ backbone is shielded by the SPG sheath, by which unfavorable interpolymer stacking of the PT-N$^+$ backbone is strongly restricted even in the film. In addition, although the SPG sheath surrounds the PT-N$^+$ backbone, the characteristic wrapping modes, e.g., somewhat elongated helical pitch, would allow small molecules to interact with the PT-N$^+$ backbone directly, as observed in doping of the PANI/SPG composite. This is in contrast to the sheath of CD arrays, providing a cyclic closed cavity.

The CD spectra of the composites show an intense split-type ICD in the π–π^* transition region (Fig. 17c). This fact clearly suggests that PT-N$^+$ would be chirally twisted in an intrastranded manner. The observed ICD patterns are characteristic of a right-handed helix of the PT backbones, reflecting the

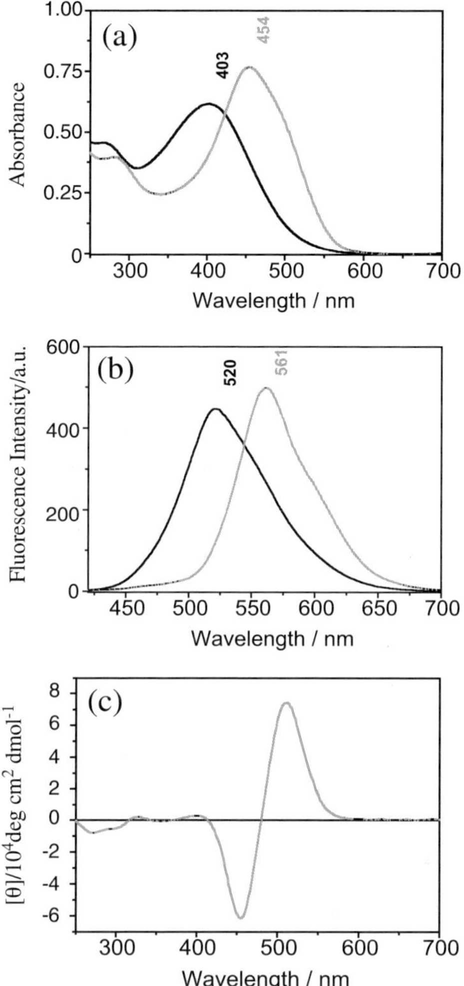

Fig. 17 a UV-VIS and **b** Emission (λ_{ex} = 400 nm) spectra of PT-N$^+$: the spectra were taken without SPG (*black line*) and with SPG (*gray line*)

right-handed helical structure of SPG [105]. The stoichiometry of the complex was determined by means of continuous-variation plots (job plots) from the CD spectra (Fig. 18). Consequently, the molar ratio of glucose residues along the main chain to the repeating unit of PT-N$^+$ can be determined to be around 2. The complexation mode of the PT-N$^+$/SPG complex is almost similar to that of the polynucleotide/SPG complex, leading to the conclusion that SPG tends to form hetero-macromolecular complexes by exchanging one glucan chain in t-SPG for one water-soluble guest polymer.

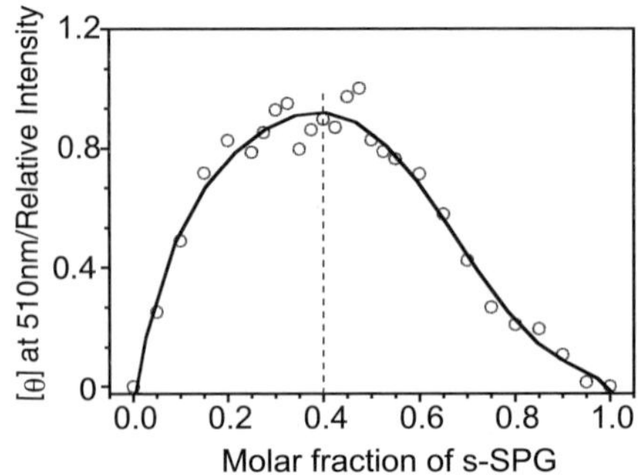

Fig. 18 Job plot obtained from the CD spectra

2.2.3
Inclusion of Poly(p-phenyleneethynylene) (PPE)

PPE is a conjugated polymer which can convert chemical and physiological signals into optical emission with signal amplification. Therefore, PPE has been regarded as a conjugated polymer suitable for chemosensors due to its excellent optical response to environmental variation through the relatively free rotation of the alkenyl-aryl single bond [106–113]. A particularly challenging aspect of PPE is to design a water-soluble PPE backbone with a one-handed helical structure because it is readily applicable as a sensitive chiral sensor targeting biologically important molecules and polymers. As a novel approach toward the creation of a chiral insulated PPE wire, β-1,3-glucans should exert their unique hosting abilities [114].

Aiming at the creation of a well-characterized complex, water-soluble PPE (PPE-SO$_3^-$) was used in the present study. Expectedly, upon mixing of PPE-SO$_3^-$ and s-SPG, a clear solution could be obtained. The resultant solution was characterized by UV-VIS, fluorescence and CD spectroscopies. Figure 19 compares the absorption spectra between PPE-SO$_3^-$ itself and its mixture with s-SPG. The absorption maximum of 442 nm observed in the absence of SPG is attributed to a random-coiled conformation of the PPE backbone. Upon mixing with s-SPG, however, the absorption maximum is red-shifted to 470 nm, indicating that the effective conjugation length of the PPE backbone is increased. The shift clearly shows that the interpolymer interaction between PPE-SO$_3^-$ and s-SPG would force the PPE backbone to adopt a planer and rigid conformation. Furthermore, fluorescence spectra of the PPE-SO$_3^-$/SPG composite revealed that the emission intensity dramatically increases upon addition of s-SPG (Fig. 20). The finding indicates that the PPE backbones do not aggregate by themselves but become more isolated through the complexation with s-SPG, supporting the view that the PPE backbone is insulated into the one-dimensional SPG cavity.

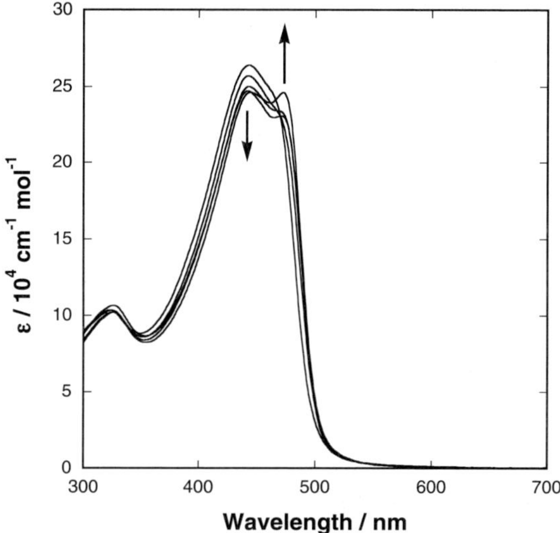

Fig. 19 UV-VIS spectral change of PPE-SO$_3^-$ as a function of SPG concentration (concentration range of s-SPG : 0–6.7 × 10^{-4} M). The concentration of PPE-SO$_3^-$ was kept at 1.5 × 10^{-4} M. H$_2$O/DMSO = 95/5 (v/v), 1.0 cm cell length, room temperature. Reprinted with permission from [114]

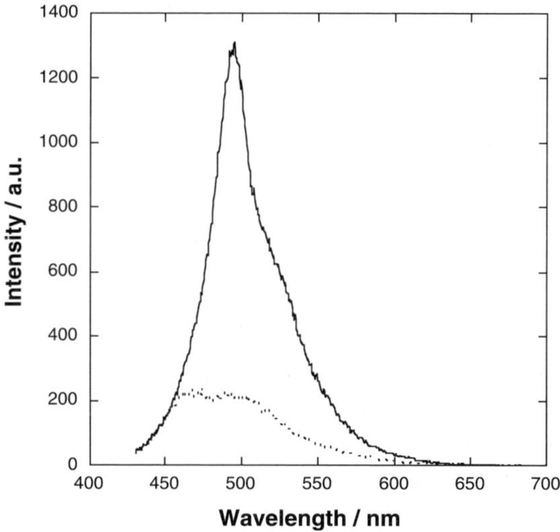

Fig. 20 Emission spectra of PPE-SO$_3^-$ in the absence of s-SPG (*dashed line*) and in the presence of 15.0 eq. (2.3 × 10^{-3} M) of s-SPG (*solid line*), λ_{ex} = 400 nm. The concentration of PPE-SO$_3^-$ was kept at 1.5 × 10^{-4} M. H$_2$O/DMSO = 95/5 (v/v), room temperature. Reprinted with permission from [114]

CD spectroscopic studies are also helpful to investigate the conformational changes of incorporated $PPE\text{-}SO_3^-$. The shape and ICD pattern of the composite suggest that $PPE\text{-}SO_3^-$ would adopt a right-handed helix which is transcribed from SPG to the PPE backbone (Fig. 21). As a reference experiment, we mixed an aqueous solution containing t-SPG with $PPE\text{-}SO_3^-$. However, the mixture did not give any CD signal around the same wavelength region. This result supports the view that the renaturating process from s-SPG is indispensable for the effective interaction between SPG and $PPE\text{-}SO_3^-$. The spectroscopic results above mentioned lead to the conclusion that $PPE\text{-}SO_3^-$ is incorporated into the one-dimensional SPG cavity as an individual polymer and its backbone adopts a right-handed twisted conformation along the SPG polymer chain. These results are almost consistent with those observed for the $PT\text{-}N^+$/SPG system, encouraging us to pursue the fundamental properties of SPG as a one-dimensional host.

To elucidate the mechanism for $PPE\text{-}SO_3^-$/SPG complex formation, the stoichiometry of the complex was determined by means of a continuous-variation plot (Job plot) from its CD spectroscopic change. From the Job plot, the maximum complex formation is attained at around 2, which corresponds to the molar ratio of the glucose residue along the s-SPG main chain to the repeating unit of $PPE\text{-}SO_3^-$. Furthermore, considering the facts that t-SPG forms a right-handed 6_1 triple helix with a 1.8 nm pitch and that three p-phenylene ethynylene units have statistically 1.6 nm length in average, we suppose that the $PPE\text{-}SO_3^-$/SPG complex prepared from a mixture of s-SPG and $PPE\text{-}SO_3^-$

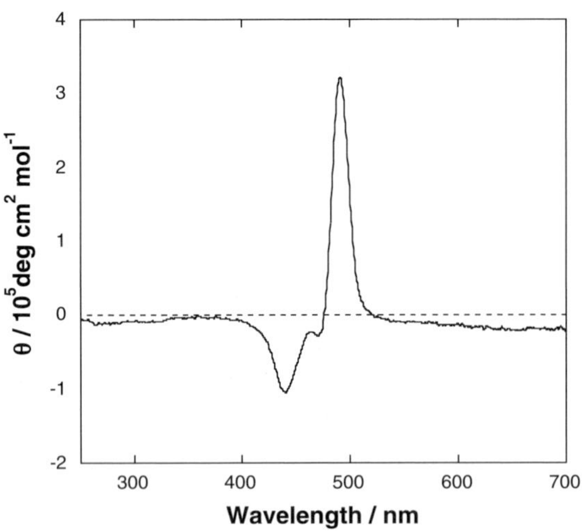

Fig. 21 CD spectra of $PPE\text{-}SO_3^-$ in the absence of 15.0 eq. (2.3×10^{-3} M) of s-SPG (*dashed line*) and in the presence of $PPE\text{-}SO_3^-$ (*solid line*). The concentration of $PPE\text{-}SO_3^-$ was kept at 1.5×10^{-4} M. H_2O/DMSO = 95/5 (v/v), 1.0 cm cell length, room temperature. Reprinted with permission from [114]

is constructed by two s-SPG chains and one PPE-SO$_3^-$ chain in which the two s-SPG chains force the PPE-SO$_3^-$ chain to be twisted. Together with the results obtained from polynucleotide/SPG and PT/SPG complexes, it can be concluded that when s-SPG interacts with relatively hydrophilic guest polymers, the resultant composites are always composed of two SPG polymers and one guest polymer.

2.2.4
Inclusion of Permethyldecasilane (PMDS)

Oligosilanes have been investigated as attractive functional materials, since they have unique σ-conjugated helical structures and show unique optoelectric properties [115–117]. Their fabrication using the one-dimensional cavity of SPG would be useful for various applications, such as (1) preparation of water-soluble oligosilane composites, (2) regulation of their helical conformation and (3) fabrication of their fibrous superstructures, etc. We have demonstrated that permethyldecasilane (PMDS, structure see: Fig. 3) is incorporated into the SPG cavity through our successful procedure to a one-dimensional composite, in which PMDS is indeed included inside the one-dimensional cavity of SPG [118]. Several lines of evidence including UV-VIS, CD and fluorescence spectroscopic data along with observations using a TEM and AFM have clearly revealed that water-soluble, helically ordered oligosilane-nanofibers are formed with s-SPG through the renaturation process.

Unlike the forgoing guest polymers, i.e., SWNT, PANI, PT, and PPE, all of which are soluble or dispersible either into water or into polar organic solvents, PMDS is soluble only in nonpolar organic solvents such as hexane. Accordingly, to prepare highly ordered PMDS/SPG composites with excellent reproducibility, establishment of a novel solubilization strategy is desired. A new biphasic procedure, on the basis of the renaturation process of s-SPG on the liquid/liquid interface, is thus exploited.

The triple strand of SPG is dissociated into the single strand at pH > 12, whereas it retrieves the original triple strand by pH neutralization [51]. Accordingly, in the present study, a NaOH solution containing s-SPG was neutralized by acetic acid, where the renaturing from s-SPG to t-SPG proceeds with decreasing pH values. Firstly, a hexane layer containing PMDS and an aqueous NaOH layer containing s-SPG were well homogenized by sonication. Aqueous acetic acid was then added to the resultant mixture to give two layers, where the renaturing from s-SPG to t-SPG would occur on the H$_2$O/hexane interface. The hexane layer thus obtained showed a broad UV-VIS absorption band with full-width-half-maximum (FWHM) of 40 nm around 280 nm, which is characteristic of the σ–σ^* transition of flexible, random-coiled PMDS. Detailed investigation of the concentration dependency of s-SPG indicated that the intensity at 280 nm in the hexane layer

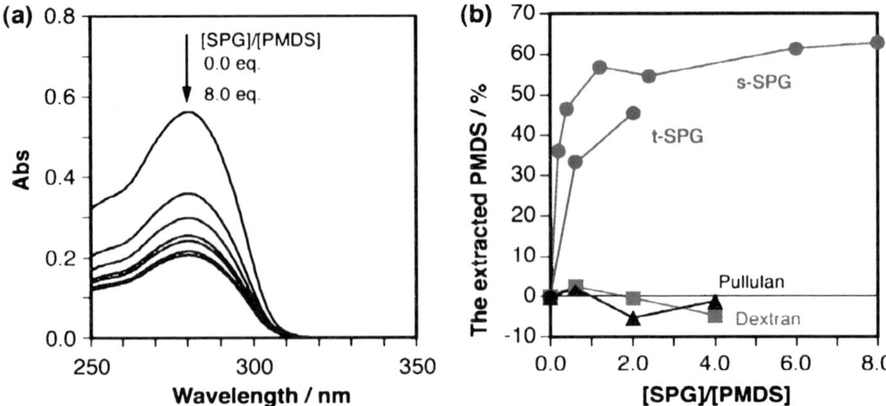

Fig. 22 **a** UV absorption spectra of the hexane layer after the treatment with aqueous layers containing different amounts of s-SPG and **b** Estimated percentages of PMDS extracted from the hexane layer into the aqueous one: pathlength 5 mm, 25 °C, [PMDS] = 0.431 mM (an original concentration in the hexane layers), [s-SPG] = 0.0–3.5 mM (0.0–8.0 eq., an original concentration in the aqueous layers) based on their repeating units (four glucoside units for SPG and one dimethylsilane unit for PMDS). Reprinted with permission from [118]

was diminished with increasing s-SPG concentration in the aqueous layer, where PMDS should be extracted into the aqueous layer with accompanying PMDS/SPG composite formation. The extraction efficiency reached ca. 60% at the highest s-SPG concentration (Fig. 22). The PMDS/SPG composite in a neutral aqueous solution showed an intense red-shifted fluorescence band with FWHM of 18 nm at 323 nm when excited at the σ–σ^* transition band (280 nm), compared to the corresponding fluorescence band at 310 nm of free PMDS in hexane.

Although the resultant aqueous layer showed the broad absorption band around 280 nm, no detectable CD signal was observed at this wavelength, suggesting that PMDS adopts a flexible, random-coiled conformation in the one-dimensional SPG cavity. From ^{1}H NMR analysis of the aqueous layer, i.e., in D_2O, PMDS and hexane are assumed to be co-included within the one-dimensional SPG cavity: in other words, PMDS is wetted by co-extracted hexane molecules (Fig. 23). Once the hexane layer was removed by lyophilization of the aqueous layer, followed by re-dissolution of the resultant PMDS/SPG composite into water, the resultant aqueous solution showed a sharp absorption band at 290 nm with a narrow FWHM of 12 nm as well as a positive CD signal at 283 nm and a negative one at 293 nm (Fig. 24). The small FWHM of this UV band suggests a very rigid, extended conformation of PMDS in the PMDS/SPG [119].

So far, detailed conformational studies on PMDS itself have been performed by using UV-VIS and CD spectroscopies. Accordingly, on the basis

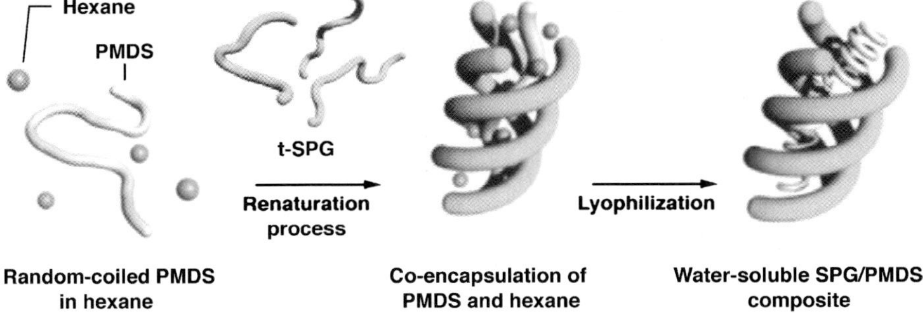

Fig. 23 Schematic illustration of the composite formation of s-SPG with PMDS. Reprinted with permission from [118]

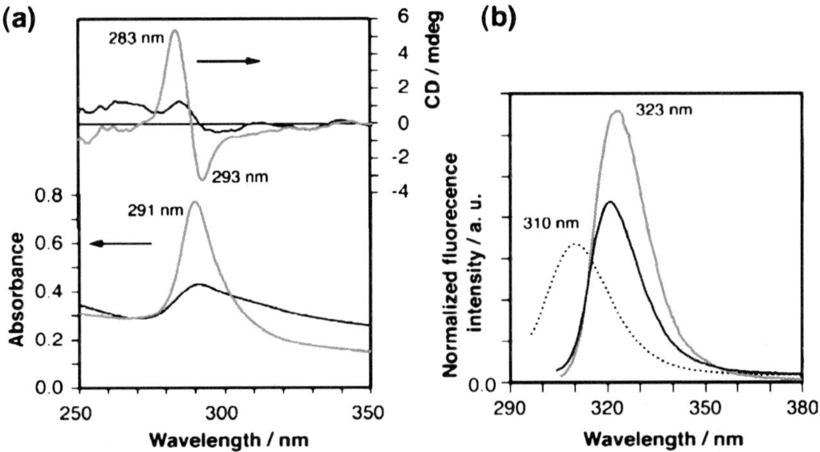

Fig. 24 **a** UV (*bottom*) and CD (*top*) and **b** Fluorescence spectra of the PMDS/SPG composite in water (λ_{ex} = 290 nm, *gray solid line*), the PMDS/t-SPG mixture in water (λ_{ex} = 290 nm, *black solid line*) and free PMDS in hexane (λ_{ex} = 280 nm, *black dotted line*): [s-SPG]/[PDMS] = 1.2 in molar ratio, cell length 0.5 cm and 25 °C. Reprinted with permission from [118]

of these reported results, it should be worth to compare the spectral feature of PMDS/SPG composite with PMDS itself, because the comparison would allow us to investigate how PMDS chain is wrapped by SPG. The origin of the bisignate CD spectral profile indicates two possibilities: (1) a mixture of two different helices with the opposite screw senses and different pitches and (2) exciton couplet due to chirally-twisted PMDS aggregates. Kunn's dissymmetry ratio being defined as $g_{abs} = \Delta\varepsilon/\varepsilon = $ CD (in mdeg)/32 980/Abs.), which is a dimensionless parameter to semi-quantitatively characterize helical structures of oligosilanes and other chromophoric chiral molecules, excludes the latter case, because the evaluated g_{abs} values at two extrema are

$+ 3.3 \times 10^{-4}$ at 283 nm and $- 1.7 \times 10^{-4}$ at 293 nm, respectively. The absolute magnitude in these small g_{abs} values is almost comparable with the g_{abs} value of $(2.0–2.5) \times 10^{-4}$ at 323 nm of rigid rod-like poly(silane) (PS) with a single-screw helix. PMDS incorporated in the helical cavity of s-SPG, therefore, exists as a mixture of diastereomeric helices with the opposite screw senses: the 283 nm CD signal is responsible for a P-7_3 helix and the 293 nm CD signal is for an M-15_7 helix. Additionally, UV-VIS and CD spectral features of PMDS/SPG composite are very similar to those for the PMDS/γCD composite [77].

2.3
Inclusion of Supramolecular Dye Assemblies

Having established the fundamental hosting abilities toward functional polymers, the one-dimensional hosting system was further extended to small molecular guests which possess potential abilities to self-assemble into one-dimensional supramolecular architectures. So far, creation of well-regulated supramolecular assemblies from a rationally designed dye molecule have attached the wide-spread interest in view of their potential applications for nonlinear optical and photorefractive devices [120–124]. A particularly challenging aspect is to create a wide variety of supramolecular assemblies from a single dye by using an appropriate template, on which dye molecules can be arranged in a well-regulated fashion to generate a wide variety of colors as well as functions, reflecting the higher-order structures of the template [10, 11, 19, 125–138].

A dipolar dye (azo-dye) having pyridine and carboxylic acid terminals has potential self-assembling capabilities through intermolecular interactions in addition to cooperative π–π stacking and dipolar–dipolar interactions [139, 140]. To achieve the different molecular arrangement of azo-dye molecule in the presence of SPG template, the different renaturing solvents, e.g., DMSO or NaOH solution were employed: in DMSO solution a self-assembling structure of azo-dye would be more dominated by the hydrogen-bonding interaction, whereas in NaOH solution π–π stacking and dipolar–dipolar interactions in addition to hydrophobic interactions would become major driving forces (Fig. 25).

As a first experiment, DMSO solution containing s-SPG was mixed with a DMSO solution of azo-dye followed by addition of water allowing renaturing to proceed. UV-VIS spectra of the azo-dye/SPG solution thus obtained was compared with azo-dye itself in the same solvent. The absorption maximum of azo-dye itself appears at 446 nm, whereas the peak of the composite is red-shifted to 468 nm [141]. This red-shift can be ascribed to the formation of J-type assemblies, promoted by the intermolecular hydrogen bonding in addition to π–π stacking interactions among azo-dye molecules. When a DMSO solution containing the azo-dye itself was diluted with water, the

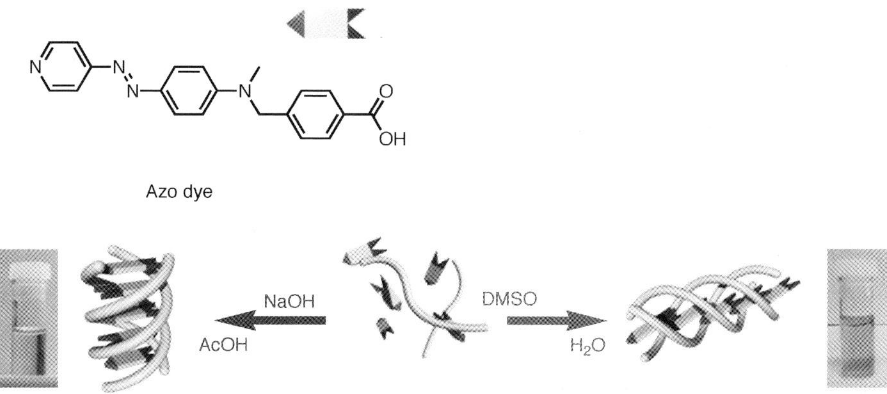

Fig. 25 Concept of the creation of supramolecular dye architectures showing "polymorphism". Structure of the designed dipolar dye with pyridine and carboxylic acid terminals. Reprinted with permission from [141]

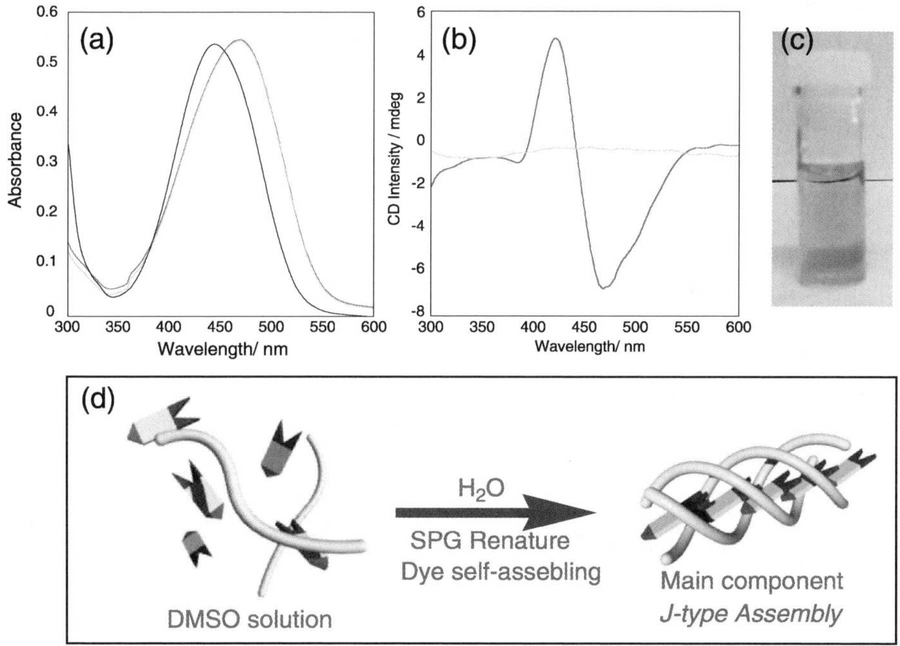

Fig. 26 **a** UV-VIS and **b** CD spectra of the samples containing azo dye/SPG composite (*black dotted lines*), monomeric azo dye in DMSO (*black line*) and azo dye aggregate in water/DMSO mixed solvent (*gray lines*), prepared from DMSO solution, 1.0 cm cell length, room temperature, **c** Photo image of the solution containing azo dye/SPG composite and **d** Schematic illustration of the J-type assembly formation during the renature of s-SPG: This type of assembly would be created as a major component in the solution. Reprinted with permission from [141]

mixture provided the precipitate after several minutes. When UV-VIS spectra of the temporarily "transparent" solution were taken immediately after addition of water, the absorption maximum appeared at 468 nm. These results support the view that a J-type assembly is entrapped into the SPG cavity. CD spectroscopy is helpful for monitoring the definitive interaction between SPG and azo-dye. Upon mixing with s-SPG an intense split-type ICD appeared in the π–π^* transition region of the azo-dye assembly. The self-assembling nanofiber structure of azo-dye is entrapped into the helical SPG cavity, adopting a twisted molecular arrangement (Fig. 26).

The triple strand of SPG is dissociated into the single strand at pH > 12, whereas it retrieves the original triple strand by pH neutralization. As an alternative strategy, we tried to construct a different type of azo-dye assemblies utilizing this neutralization process of the alkaline s-SPG solution, where π–π stacking and dipolar–dipolar interactions become the major driving forces instead of hydrogen-bonding interactions. To the NaOH solution containing

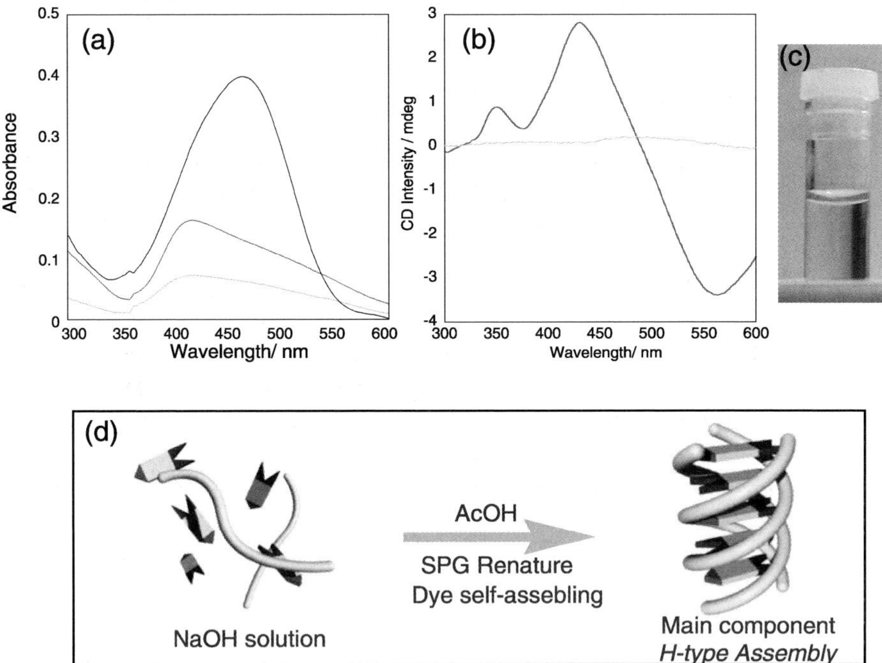

Fig. 27 **a** UV-VIS and **b** CD spectra of the samples containing azo dye/SPG composite (*black dotted lines*), monomeric azo dye in DMSO (*black line*) and azo dye aggregate in water/DMSO mixed solvent (*gray lines*), prepared from NaOH solution, 1.0 cm cell length, room temperature, **c** Photo image of the solution containing azo dye/SPG composite and **d** Schematic illustration of the H-type assembly formation during the renaturation of s-SPG: This type of assembly would be created as a major component in the solution. Reprinted with permission from [141]

s-SPG and azo-dye, aqueous acetic acid was gradually added to give a clear yellow solution, adjusting the final pH to 7. The absorption maximum of the solution is blue-shifted from 446 to 417 nm with accompanying slight peak broadening. This blue-shift is ascribed to the creation of the H-type assembly. Furthermore, ICD is also detected at the $\pi-\pi^*$ transition region of the azo-dye assembly, indicating that the one-dimensional H-type assembly is entrapped in the helical SPG cavity, where $\pi-\pi$ stacking and dipolar–dipolar interactions in addition to hydrophobic interactions would become the major driving forces (Fig. 27). It should be noted that the entrapped supramolecular structures of azo-dye are not affected by the solvent properties, meaning that the observed color changes are not due to the difference in solvent properties surrounding the composites. These results clearly support the view that the different dye assemblies can be created from azo-dye through the different preparation procedures. One may regard this phenomenon to be a sort of "polymorphism" induced by the presence of SPG. The creation of similar dye assemblies has been achieved by using porphyrin derivative, where a J-type porphyrin assembly is also entrapped into the helical SPG cavity [142].

2.4
β-1,3-Glucan as a One-Dimensional Reaction Vessel

Our research efforts are particularly focused not only on polymeric guests but also on low molecular weight compounds. Especially, it is of great significance to establish β-1,3-glucan-templated polymerization of various monomers in the one-dimensional cavity to construct the corresponding polymers with fibrous morphologies, where SPG or CUR is expected to act not only as a one-dimensional host for monomers but also as a one-dimensional vessel for the stereoselective polymerization reaction (Fig. 28). Herein, we show several successful examples where SPG accommodates various reactive monomers, including 1,4-diphenylbutadiyne (DPB) derivatives, EDOT and alkoxysilane, within the one-dimensional cavity and produces unique water-soluble nanofibers through in situ polymerization reactions.

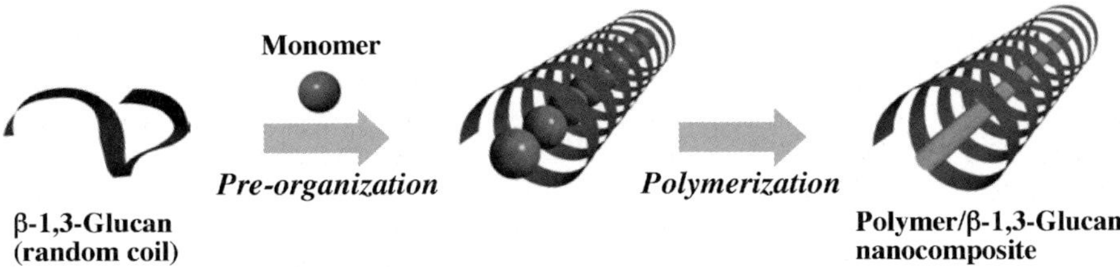

Fig. 28 Concept of in situ polymerization reaction utilizing β-1,3-glucan as a vessel

2.4.1
Photo-Polymerization of Diacetylene Derivatives

Poly(diacetylene)s are known as a family of the most interesting research targets among the π-conjugated polymers, because they are readily produced through photo-irradiation (UV or γ-ray) without any initiators [143–145]. Instead, for the photo-mediated polymerization of diacetylene derivatives, closely packed pre-organization of the corresponding monomers is indispensable [146–148]. Therefore, this topochemical polymerization process of diacetylene derivatives was mainly studied only in crystals, micelles and Langmuir–Blodgett films, where corresponding monomers are aligned in a parallel but slightly slided packing mode. In the present study, 1,4-diphenylbutadiyne (DPB) derivative was used as a monomer and pre-organization of DPB can be easily achieved by incorporation into the one-dimensional SPG cavity [149, 150].

The general procedures are as follows: a DMSO solution containing s-SPG was mixed with DPB DMSO solution and the resultant mixture was diluted with water to regenerate a t-SPG helical structure. Although DPB was scarcely soluble by itself in water, the resultant slightly turbid solution showed a clear CD spectrum, which is assignable to an absorption band of DPB. As reference experiments, when other polysaccharides and carbohydrate-appended detergents (amylose, pullulan, dextran, and dodecyl-β-D-glucopyranoside) were used instead of s-SPG, no or a negligibly weak CD signal was observed (Fig. 29). Together with the fact that SPG or DPB itself gave no CD signal at this wavelength region, the observed negative CD exciton-coupling is indicative of twisted-conformations or -packings of DPB arising from the strong helix-forming capability of SPG. These results support the view that DPBs are pre-organized only in the presence of s-SPG to form a one-dimensional supramolecular assembly, which is favorable to subsequent polymerization reaction under UV-irradiation.

UV-irradiation using a high-pressure Hg-lamp induced a gradual color change of the solution containing the DPB/SPG complex from colorless to pale blue. The UV-VIS spectrum of the resultant solution shows an absorption band at around 720 nm which is characteristic of poly(diacetylene)s with extremely long π-conjugated length and/or tight inter-stranded packing [151, 152]. UV-mediated polymerization of DPB was also confirmed by Raman spectra, in which the DPB/SPG complex shows a clear peak at $2000\,\text{cm}^{-1}$ assignable to poly(diacetylene)s (–CH=CH– stretching vibration) after 16 h UV-irradiation. On the contrary, no such Raman peak appeared without SPG. Together with the fact that UV-mediated polymerization of diacetylenes proceeds in a topochemical manner, these data suggest that SPG accommodates DPB to align them in a packing suitable for such topochemical polymerization. It should be noted that p-amido-functionalities of DPB are essential for the UV-mediated polymerization. We assume that the p-amido-

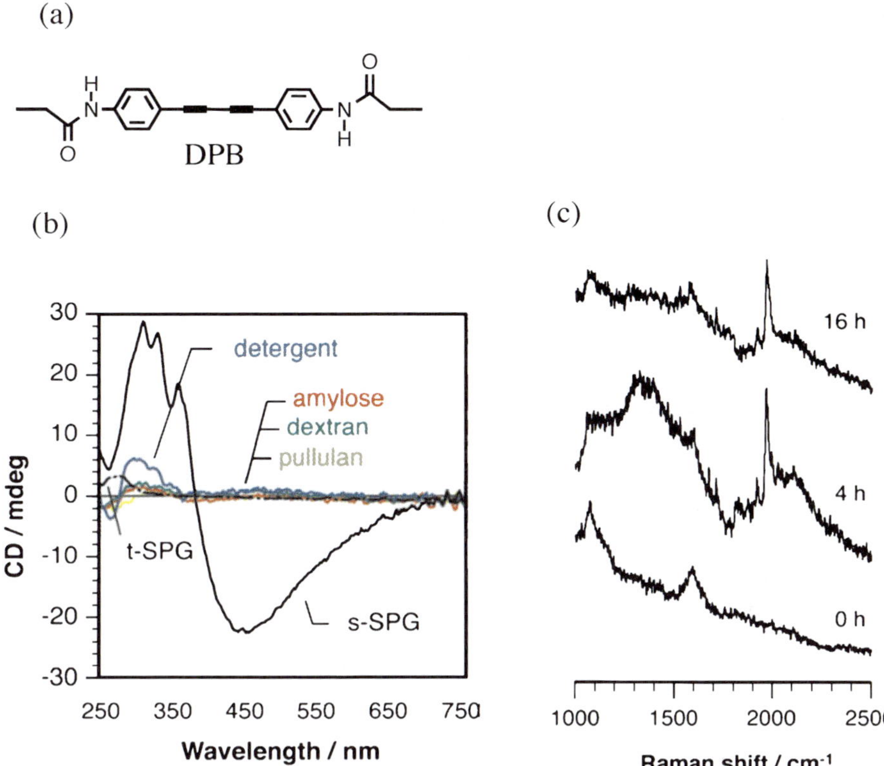

Fig. 29 a Structure of 1,4-diphenylbutadiyne (DPB) derivatives, **b** CD spectra of DPB in the presence of s-SPG, amylose, dextran, pullulan, t-SPG and the detergent: cell length 1.0 cm, 20 °C, [H$_2$O] = 70% v/v, [DPB] = 25 µg/ml, [polysaccharide] or [detergent] = 25 µg/ml, **c** Raman spectra of DPB in the presence of s-SPG after 0, 4 and 16 h

functionalities should form hydrogen-bonds with SPG and/or neighboring monomers to orientate the monomers in suitable packing arrangements for the polymerization. The results presented here will open a way to construct poly(diacetylene)-nanofibers with uniform diameters from various DPB-derivatives.

2.4.2
Sol–Gel Polycondensation Reaction of Alkoxysilane

Creation of supramolecular organic/inorganic hybrid materials has been of great concern in recent years [153–155]. In particular, inorganic nanofibers which have a well-regulated shape and high water-solubility, reflecting those of organic templates, are desired for potential applications to biosensors, switches, memories, and circuits [156–159]. So far, silica nanofibers have been created using the templating method where anionic silica particles are deposited on a cationic fibrous template. However, the difficulties in surface modification, by which one may dissolve the organic/inorganic composite into the solvent, still remain unsolved because the solvophobic inorganic layer always exists outside the solvophilic organic template. It thus occurred to us that when hydrophobic metal alkoxides are entrapped in the one-dimensional SPG cavity, sol–gel polycondensation takes place inside the cavity to afford a silica nanofiber which shows the water-solubility as well as the biocompatibility arising from surface-covering SPG, making biological applications. Here, we carried out the sol–gel polycondensation reaction in the presence of s-SPG using TMPS (trimethoxypropylsilane) as a monomer [160].

TMPS was dissolved in DMSO and mixed in solution with s-SPG or s-CUR during the renaturing process. After leaving the mixed solution for 20 days at room temperature, the resultant water/DMSO mixed solvent was subjected to dialysis. The IR spectrum showed the appearance of a strong new vibration band assignable to the Si–O–Si bond at around the $1000–1200\,cm^{-1}$ region. Interestingly, the obtained silica did not form any precipitate even after dialysis against water, indicating that the created silica is successfully wrapped by hydrophilic SPG or CUR.

The aqueous silica suspension obtained from the TMPS/SPG system was cast on a grid with carbon mesh and the morphologies of the obtained silica were observed by TEM. As expected, the fine silica nanofiber structure with a uniform 15 nm diameter is observed (Fig. 30), whereas no such fibrous structure is found from the sample prepared in the absence of SPG. From these TEM images, one can propose that sol–gel polycondensation of TMPS predominantly proceeds in the one-dimensional SPG cavity to afford silica nanofibers soluble in water. Furthermore, the diameter of the obtained silica nanofiber increased (ca. 25 nm) with the increase in the feed TMPS concentration up to TMPS/SPG repeating unit = 3.0 eq. However, when more than 3.0 eq. TMPS against s-SPG repeating unit was used, the silica nanofiber structure was no longer created and instead an amorphous silica mass was formed. The results support the view that under such a condition where TMPS exists in great excess, s-SPG cannot cover the original one-dimensional assembly structure of TMPS during the renaturating process, so that it can no longer act as the host effectively. The view is also supported by the fact that sol–gel polycondensation in the presence of t-SPG instead of s-SPG did not give any fibrous structure; that is, the renaturating process from s-SPG to t-SPG is indispensable for inclusion of TMPS followed by the creation of the silica nanofiber structure.

It is well-known that a catalytic amount of benzylamine or HCl allows the sol–gel polycondensation to proceed. In the present system, however, we could not find any fibrous structure from such reaction conditions. The result indicates that the very slow condensation reaction is favorable for one-dimensional growth of silica nanofibers. In fact, it took more than two weeks to create the complete nanofiber structure. Thus, to elucidate the reaction

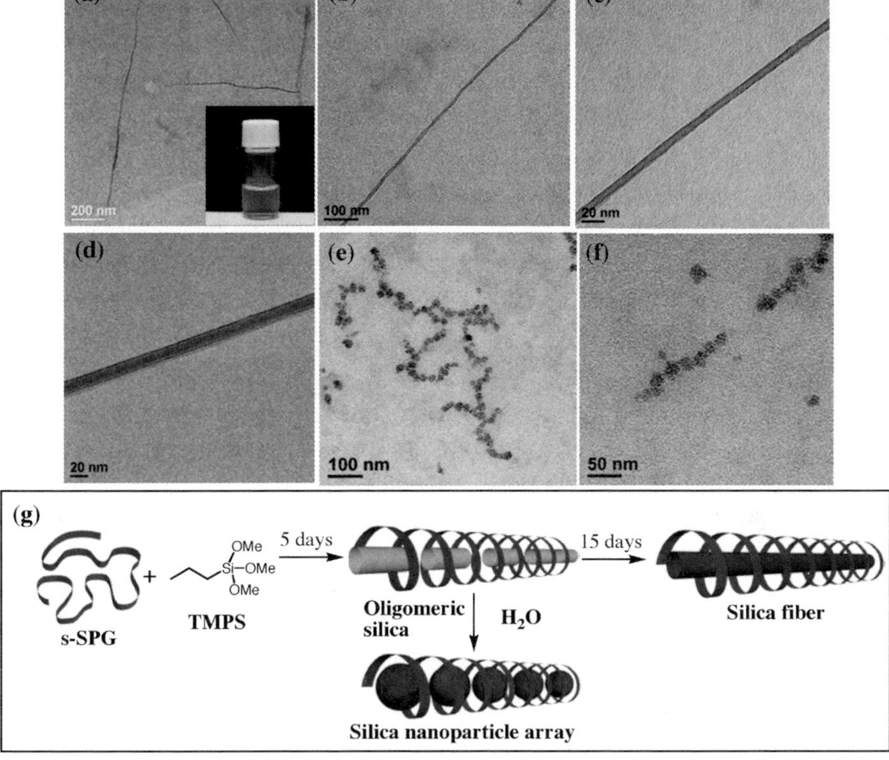

Fig. 30 a TEM image of silica nanofibers (*inset*: photo image of aqueous solution of silica nanofibers) (TMPS/SPG repeating unit = 1.0 eq.), **b, c** Its magnified images, **d** TEM image of silica nanofiber (TMPS/SPG repeating unit = 3.0 eq.), **e** TEM image of silica nanoparticle one-dimensional arrays, **f** Its magnified image and **g** Schematic illustration of our concept to utilize SPG as a vessel for sol–gel polycondensation reaction

mechanism of the polycondensation reaction, we stopped the sol–gel reaction only after 5 days and observed the morphologies of the obtained silica by TEM. Very interestingly, the morphology of the obtained silica is not a fiber but a one-dimensional array of silica nanoparticles with 10–15 nm. Here, one can propose the possible growth mechanism as the following; that is, SPG acts as a one-dimensional host for TMPS or its oligomers and the polycondensation occurs inside the cavity to give rod-like oligomeric silica after 5 days. This rod-like oligomeric silica is easily destroyed and finally converted into particles during the dialysis process due to the hydrophobic interaction among propyl groups. These results strongly support the view that SPG and CUR have a potential ability to act not only as a one-dimensional host for TMPS but also as a vessel for sol–gel polycondensation reactions. We believe that the present system would be readily applicable to the creation of novel organic/inorganic hybrid nanomaterials and their functionalized derivatives.

2.4.3
Chemical Polymerization of 3,4-Ethylenedioxythiophene (EDOT)

Poly(3,4-ethylenedioxythiophene) (PEDOT) has been widely investigated during the past decade owing to its low band gap, high conductivity, good environmental stability, and excellent transparency in its oxidized state [161–164]. More recently, the preparation and characterization of PEDOT nanostructures have become a topic of increasing interest due to their potential applications to electrical, optical, and sensor devices [165–169]. On the basis of the established one-dimensional hosting abilities of β-1,3-glucans, herein, we carried out a polymerization reaction of EDOT in the presence of s-SPG initiated by the oxidant ammonium persulfate (APS), expecting that PEDOT nanofibers are created in the SPG cavity [170]. After the polymerization reaction, homogeneous and stable aqueous dispersions with a dark blue color were obtained. Subsequently, the aggregated structure of the PEDOT/SPG composites was examined by TEM (Fig. 31). Unexpectedly, it is clearly seen from the TEM images that the resultant composite does not construct PEDOT nanofibers but PEDOT nanoparticles with uniform size and regular shape. Interestingly, when the concentration of used SPG increased, the diameter of the nanoparticles decreased from 160 nm ([s-SPG] = 0.5 mg/ml) to 70 nm ([s-SPG] = 3.0 mg/ml), indicating that SPG acts as a template for the polymerization reaction. Further reliable evidence that s-SPG really acts as the template for the particle formation was obtained from the energy dispersive X-ray spectroscopy (EDS); that is, the O/S elements ratio in the nanoparticle clearly shows that PEDOT and SPG coexist in the created nanoparticles.

To elucidate the growth mechanism of PEDOT/SPG nanostructures, CD spectra of the resultant solution were recorded. The CD spectra revealed that PEDOT/SPG nanoparticles show no CD signal, suggesting that no effective interaction occurred between SPG and PEDOT. This fact leads us to the conclusion that at the initial stage, the polymerization of EDOT proceeds mainly in the bulk phase due to the good solubility of EDOT in water/DMSO mixed solvent. On the other hand, PEDOT is hydrophobic and insoluble in water/DMSO mixed solvent. Therefore, once PEDOT is formed, the polymer chains tend to aggregate through the strong intra- or intermolecular π-stacking. Consequently, the hydrophobic EDOT oligomer or polymeric PEDOT with the entangled polymer backbone may interact with s-SPG to form the PEDOT/SPG complex, where SPG cannot act as an effective one-dimensional template for such globular-shaped polymeric guests. It is worth mentioning that the PEDOT/SPG composite would act as a sort of amphiphilic block copolymer, which can self-aggregate into nanoparticles under the experimental conditions. This unexpected result would provide novel methodologies to create water-soluble PEDOT nanoarchitectures.

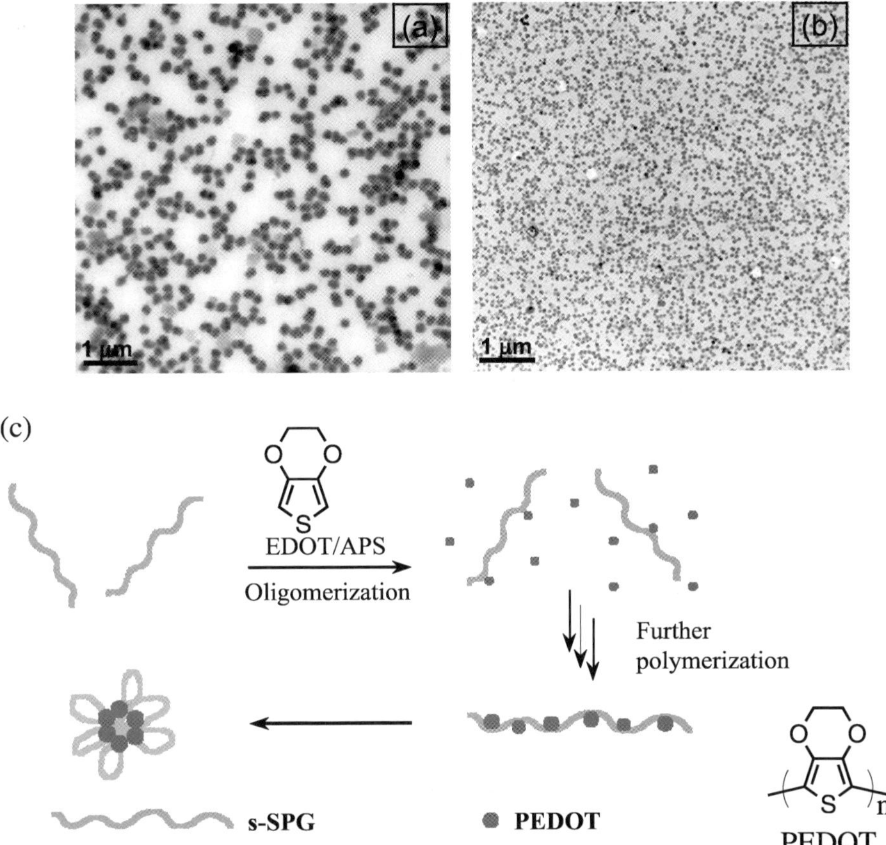

Fig. 31 TEM images of PEDOT/SPG nanoparticles prepared by APS oxidant in the presence of s-SPG ([EDOT] = 9.4 mM, [APS] = 94 mM): **a** [SPG] = 0.5 mg/ml, **b** [SPG] = 3.0 mg/ml and **c** Schematic illustration of the possible growth mechanism of water-soluble PEDOT/SPG nanocomposites

2.5
One-Dimensional Arrangement of Au-Nanoparticles

Nanoscale assemblies of inorganic nanoparticles, consisting of a limited number of metal or semi-conductive nanoparticles, have received considerable attention in recent years due to their application as nano- and biomaterials, e.g., sensors, memory, imaging agents, etc. [171–174]. Such inorganic nanoparticles, however, lack the potential abilities to arrange along specific directions as atoms or organic hybrids. Control of the self-assembling behavior of nanoparticles is, therefore, a still challenging research subject and exploitation of the versatile strategy is strongly desired [175].

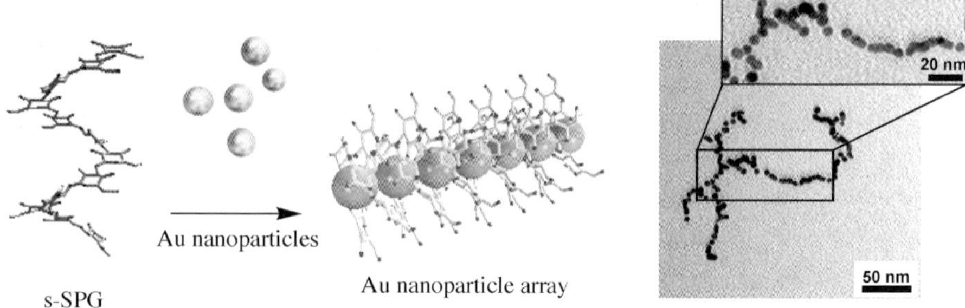

Fig. 32 TEM images of Au nanoparticle arrays

The unique renaturing process during the guest encapsulations provides one great advantage of SPG, that is, a flexible, induced-fit-type size/shape-selectivity for the guests, independent of the surface nature of the guests. This advantage of SPG is clearly demonstrated by using Au nanoparticles as a guest [176]. For example, stable Au/SPG composites were obtained from commercially available Au nanoparticles (5–50 nm). The resultant Au/SPG composites are well-soluble in water but show a significantly broadened plasmon absorption, indicating SPG-induced assemblies of Au nanoparticles in aqueous media. TEM observations of the resultant Au/SPG composites showed unique one-dimensional Au nanoarrays with their length consistent with that of s-SPG itself. These data clearly indicate that the one-dimensional nanoarrays of Au nanoparticles arise from their encapsulation within the one-dimensional cavity of SPG (Fig. 32).

SPG will act not only as a one-dimensional template for preorganization of Au nanoparticles but also for facilitation of fusion of Au nanoparticles [177].

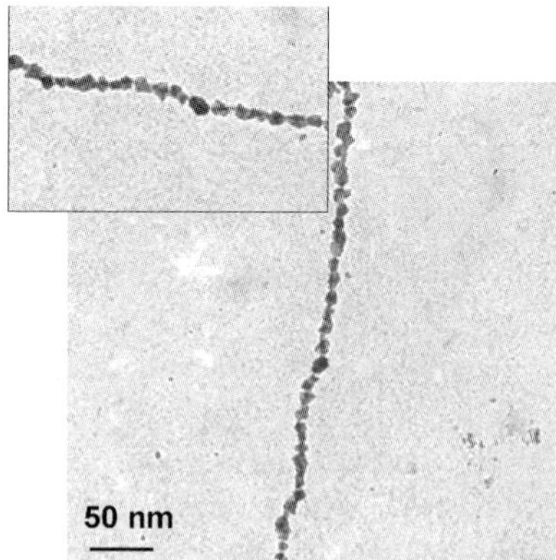

Fig. 33 TEM images showing formation of continuous nanowires: Au nanowires created using chemical reduction of HAuCl$_4$ after pulsed-laser irradiation in aqueous solution at 25 °C; 5 min, 5 mJ/pulse, 532 nm

We have found that a morphological change of one-dimensionally aligned SPG-Au composite is induced by 532 nm laser irradiation, which leads to the fusion of Au nanoparticles to prepare discontinuous Au nanowires. Furthermore, using both the near-IR irradiation and a reducing agent lead to continuous Au nanowires. This capability opens up new possibilities not only for novel sensor systems but also for applicant electronics, for example, patterning the metallization in solution for nanocircuits, which was previously very difficult (Fig. 33).

3
Chemically Modified β-1,3-Glucans

3.1
Synthetic Strategies Toward the Selective Modification

Many polysaccharide researches have focused on exploiting the functional material thorough chemical modification of polysaccharides. The problems in chemical modification of natural polysaccharide arise from the similar reactivity of the hydroxy (OH) groups, making selective functionalization difficult. Accordingly, to introduce functional groups into the desired OH groups, much effort has been paid to exploiting an effective synthetic route, facilitating access to various polysaccharide-based nanomaterials.

Chemical modification of β-1,3-glucans has been independently developed by several research groups including us [178–194]. Here, to prepare the functionalized β-1,3-glucan one-dimensional hosts, the selective modification targeting to side-glucoses for SPG and 6-OH groups for CUR need to be exploited, because 2- or 4-OH groups connected to the main-chain glucose units are indispensable for the construction of the inherent helical structure [1]. On the basis of this fact, we successfully established a versatile synthetic route to introduce various functional groups selectively into the side-glucoses of SPG and 6-OH groups of CUR, without affecting the inherent helix-forming properties of β-1,3-glucans (Schemes 1 and 2).

CUR has one primary OH group in its repeating unit, appending at C-6, i.e., the 6-OH group, which would be an active nucleophile under appropriate reaction conditions, making the selective modification of 6-OH groups possible. Moreover, the quantitative reaction can be achieved through the azideation reaction of 6-OH, followed by "click chemistry", which involves a Cu(I)-catalyzed chemoselective coupling between organic azides and terminal alkynes [192, 193]. This newly exploited strategy allows us to directly introduce various functional groups into 6-OH groups of CUR, leading to the creation of functional materials based on CUR (Scheme 2). Actually, we have successfully developed that 6-OH groups of CUR are convertible to various functional groups, e.g., ferrocene, pyrene, phorphyrin, trimethylammonium,

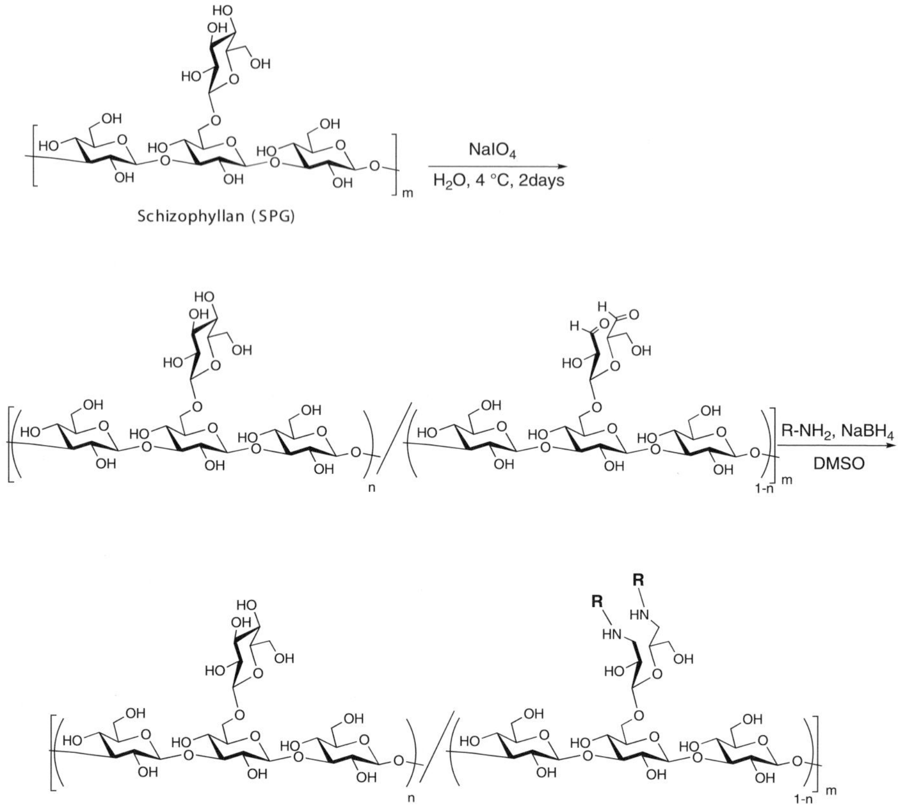

Scheme 1 Synthetic strategy for the selective modification of SPG side glucoses

sulfonate, etc., and found that these functional groups in the side chains govern the chemical properties of the modified CUR [185, 186, 192–194]. The advantageous point of this method is that a series of reactions proceed quantitatively and selectively. Additionally, by adjusting the feed acetylene composites, different functional groups are easily introduced into the same CUR chain in a step-wise manner.

SPG has two primary OH groups on the main chain and side glucose groups, so that the selective modification targeting to the primary OH groups seems to be difficult. On the other hand, the side glucose groups of SPG have 1,2-diols at 2-, 3- and 4-OH positions, whereas main chain glucoses have no such 1,2-diol due to the glicoside linkage formation between 1- and 3-OH positions. Therefore, one may expect that the oxidative cleavage of these 1,2-diols by $NaIO_4^-$ would proceed selectively only at the side glucose units. Combining this oxidative cleavage of the 1,2-diols with the reductive amination reaction, these classical synthetic strategies can be a powerful tool for the selective modification of the native SPG chain (Scheme 1). So far, we

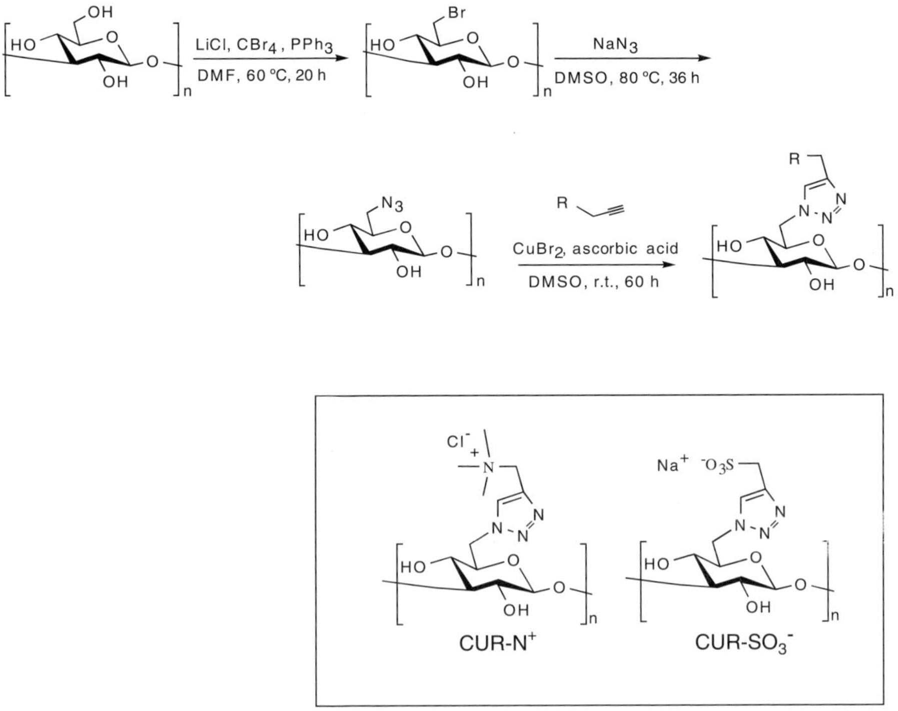

Scheme 2 Synthetic strategy for the selective modification of CUR 6-OH groups

have demonstrated that a series of chemically modified SPGs bearing various molecular recognition moieties, e.g., ionic groups, saccharide, amino acid, co-enzyme, etc., can be successfully obtained according to this synthetic strategy [184, 187–191]. Here, it should be important to address the discrepancy between these chemically modified CURs and SPGs: 6-OH groups of CUR can be converted to functional groups "quantitatively", whereas the side glucose groups of SPG can be "partially" functionalized because excessive oxidation of the side glucoses causes the insoluble aggregate probably due to the inter-polymer cross-linking between 6-OH and aldehyde groups thus formed. The modification percentage of the side glucose groups is, at most, 30%.

3.2
Partially Modified SPG: One-Dimensional Host Toward the Supramolecular Functionalization of Guest Polymers

When polymeric guests or molecules are entrapped into the SPG cavity, the side group glucose should exist on the surface of the composites. If this is the case, a functional group introduced into the side group glucose would be useful as a recognition target. To test this idea, according to the reported

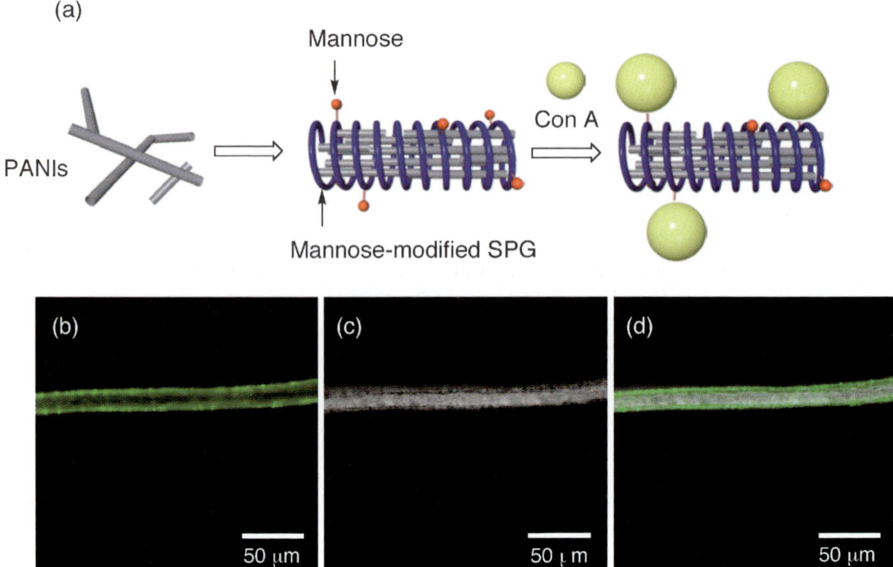

Fig. 34 **a** Concept of the supramolocular functionalization of the entrapped guest polymers. CLSM images of PANIs/mannose-modified SPG composite + FITC-ConA, **b** Fluorescence image, **c** Optical microscope image, **d** Overlap of (**b**) and (**c**). Reprinted with permission from [97]

procedure, mannose-modified SPG was synthesized and used as a wrapping reagent for PANIs [97]. It is known that this mannose group exhibits selective binding to Concanavalin A (ConA). Actually, the specific interaction between the composite and ConA was estimated by a CLSM using a FITC-labeled ConA (Fig. 34). The CLSM observation clearly shows that PANIs and ConA coexist in the same domain, indicating that (1) mannose-modified SPG can also wrap PANIs and (2) the mannose groups introduced into the side groups would exist on the exterior surface of the composite.

The findings clearly show that chemically modified SPG maintains its inherent ability as a one-dimensional host. The wrapping of the chemically modified SPG provides a novel strategy to create functional polymer composites in a supramolecular manner. Considering a general difficulty in introducing functional groups into the functional polymer backbones, the present system can be a new potential path to develop functional polymeric materials.

3.3
Semi-Artificial Polysaccharide Host Based on CUR

Even though the selective modification at 6-OH of CUR does not affect its inherent helix-forming ability, the renaturation and denaturation properties as well as the solubility of the modified CURs are strongly affected by the nature of the introduced groups, because they always exist on the helix surface. Among various functional groups introduced into 6-OH groups, so far, ionic groups, such as trimethyl ammonium and sulfonium groups are of significant interest: the electrostatic repulsion among the ionic groups on the CUR surface would provide strong influence on its conformation in water, because the ionic groups are located in the distance of approximately 6 Å if they adopt the triple-stranded helical form. Consequently, the electrostatic repulsion causes the destabilization of the triple helix, being transformed to a loosely tied triple-stranded or to a single-stranded conformation in water. This fact implies that the modified CURs have potential for acting as a one-dimensional host even without the denaturation/renaturation processes in water.

To confirm the conformational properties of CUR-N$^+$, we measured the optical rotatory dispersion (ORD) spectra at various conditions (Fig. 35A). In the form of a triple-stranded helix, the ORD spectra of β-1,3-glucan polysaccharides such as SPG and CUR have positive values at the wavelength region from 600 to 200 nm. However, CUR-N$^+$ in water shows a negative sign at this

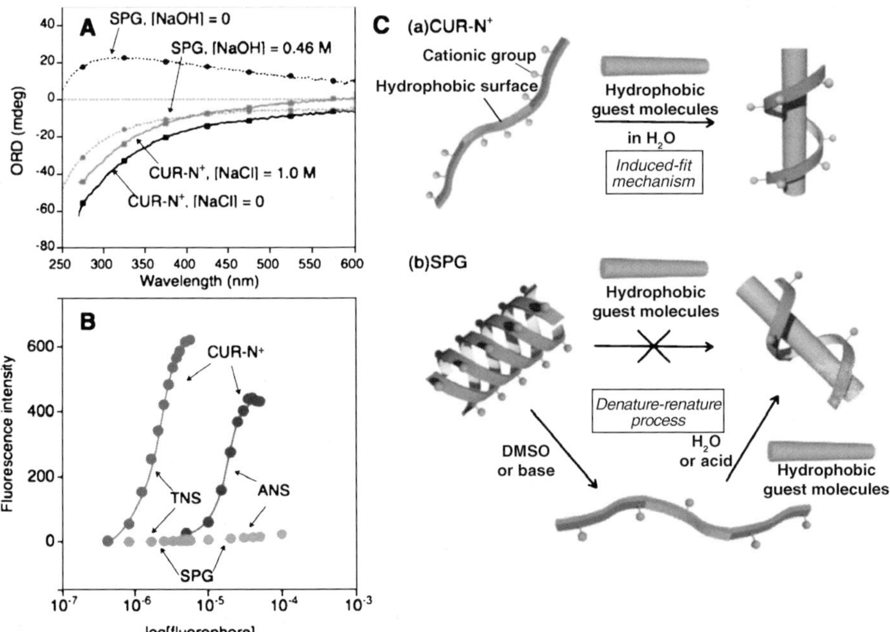

Fig. 35 A ORD curves of CUR-N$^+$ and SPG in aqueous solution (5 mg/ml) at various conditions at 25 °C, **B** Plots of fluorescence intensities at 480 nm for ANS ($\lambda_{ex} = 370$ nm) and at 450 nm for TNS ($\lambda_{ex} = 370$ nm) versus log fluorophore concentration in the presence of CUR-N$^+$ [10 μM (monomer unit)] or SPG [10 μM (monomer unit)] in aqueous solution at ambient temperature and (**C**) Schemes showing composite formation between β-1,3-glucans [(a) CUR-N$^+$ and (b) SPG] and a hydrophobic guest molecule. Reprinted with permission from [192]

wavelength region. The negative sign can be ascribed to the single-stranded form. In addition, we evaluated the effect of added NaCl on the ORD sign of CUR-N$^+$, expecting that it may reduce electrostatic repulsion. Interestingly, the intensity of ORD increased gradually with increasing NaCl concentration but did not reach the value of SPG in water. This difference implies that formation of the triple-stranded helical structure from CUR-N$^+$ is strongly suppressed by electrostatic repulsion among the cationic charges of the side chains.

These findings are successfully supported by the fluorescence probe experiments using 1-anilinonaphthalene-8-sulfonate (ANS) and 2-p-toluidyl-naphthalene-6-sulfonate (TNS) (Fig. 35B); that is, the fluorescence intensity of ANS and TNS was remarkably increased in the presence of CUR-N$^+$ or CUR-SO$_3^-$ accompanied with the blue shift of the fluorescence maxima. These results indicate that ionic CURs can accommodate ANS (or TNS) molecules into the hydrophobic domain *even in water*, where ionic CURs exert their hosting abilities in their single-stranded forms. Native SPG and CUR never exhibit such abilities in water due to their inherent nature to form the stable triple helix. SPG with cationic moieties in its side glucoses, whose modification ratios reaches 30%, also shows similar behavior in water because of its still stable triple helix. These facts lead to the conclusion that the quantitative modification of 6-OH groups with ionic groups provide a unique and unusual nature to native β-1,3-glucans [192]. The advantageous lines for CUR-N$^+$ and CUR-SO$_3^-$ as a one-dimensional host are that these CURs can act as one-dimensional hosts in water without denaturation/renaturation processes: unlike native or partially modified β-1,3-glucans, hydrophobic polymers as well as molecules can be easily entrapped into the hydrophobic domain through a simple mixing procedure (Fig. 35C).

This unusual hosting system was further applied for the polymeric guests, i.e., polycytidylic acid [poly(C)], permethyldecasilane (PMDS) and single-walled carbon nanotubes (SWNTs), which can be incorporated into SPG and CUR only through the denaturation/renaturation processes [68, 69, 84, 86, 118]. Upon just mixing these hydrophobic guest polymers with CUR-N$^+$ (or CUR-SO$_3^-$) in water, clear aqueous solutions were obtained. The characterization by UV-VIS, CD spectroscopic measurements (Fig. 36), and AFM and TEM observations revealed that these polymeric guests are entrapped into the hydrophobic domain to give stoichiometric, nanosized fibrous structures. In the case of a poly(C) guest, the complexation would be driven by the cooperative action of (1) the hydrogen-bonding interaction between the OH group at the C2 position and hydrogen-bonding sites of the cytosine ring, (2) the electrostatic interaction between the ammonium cation and the phosphate anion, and (3) the background hydrophobic interaction. Likewise, the binding properties of CUR-N$^+$ toward PMDS and SWNT are attributed to the one-dimensional amphiphilic nature of CUR-N$^+$ in water. In the case of PMDS, approximately 5% of PMDS feed can

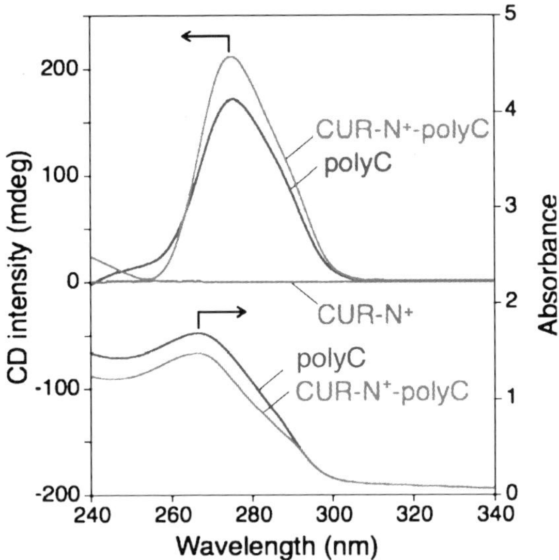

Fig. 36 UV-VIS and CD spectra of CUR-N$^+$ [1.3 mM (monomer unit)], poly(C) [0.24 mM (monomer unit)] and poly(C)/CUR-N$^+$ [1.3 mM (monomer unit), 0.24 mM (monomer unit)] in 1.0 mM tris-HCl buffer (pH 8.0) at 5 °C with a 1.0 cm cell (CUR-N$^+$ + poly(C) in absorption spectra corresponds to the sum of the individual spectra). Reprinted with permission from [192]

be solubilized into water by incorporation into the CUR-N$^+$ cavity. This value indicates that approximately 0.8 eq. of PMDS against CUR-N$^+$ in the monomer unit (SiMe$_2$ unit for PMDS and glucose unit for CUR-N$^+$, respectively) is included in PMDS/CUR-N$^+$ composites. Further conformational investigation of PMDS by CD spectroscopies reveal that the dissymmetry ratios (g) of PMDS/CUR-N$^+$ composites are smaller by a factor of ca. 4 than those of PMDS/SPG composites, indicating that the PMDS included in CUR-N$^+$ consists of a mixture of a few conformations (Fig. 37A). Together with the broad absorption band of PMDS/CUR-N$^+$ composites compared to that of PMDS/SPG composites, the strong electrostatic repulsion among cationic side-chains of CUR-N$^+$ probably prevents PDMS/CUR-N$^+$ composites from formation of a rigid supramolecular helical structure as seen in the PMDS/SPG system.

Powder of ag-SWNTs can be easily dispersed into water in the presence of CUR-N$^+$ or CUR-SO$_3^-$ with the aid of sonication, the solution color becoming dark as the sonication time progressed. The resultant aqueous solution is stable for more than one-month without forming any precipitate probably due to the electrostatic repulsion among the composites. The NIR-VIS spectral feature of the solution was similar to that of the SWNT/SPG composite solution, suggesting that CUR-N$^+$ would wrap an individual SWNT fiber in the helical cavity. The fact is also supported by TEM and AFM observations (Fig. 37B).

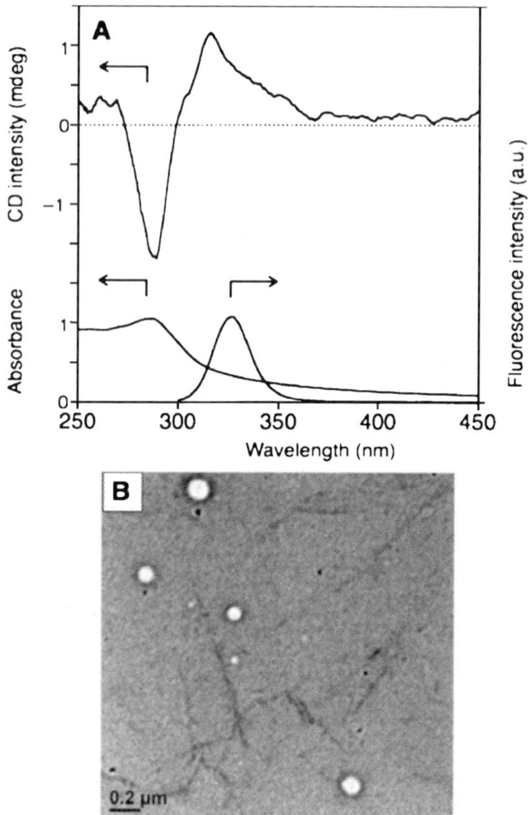

Fig. 37 A UV-VIS, fluorescence (λ_{ex} = 290 nm) and CD spectra of PMDS/CUR-N$^+$ composite ([CUR-N$^+$] = 3.0 × 10^{-4} M (monomer unit)), [PMDS] = 2.5 × 10^{-4} M (monomer unit); in an aqueous solution at ambient temperature and **B** The TEM image of PMDS/CUR-N$^+$ composite after treatment with phosphotungstic acid. Reprinted with permission from [192]

As described in the previous section, mannose-modified SPG can bestow the lectin affinity to PANI fibers by just wrapping them. The wrapping of the guest polymer by the chemically modified SPG can be regarded as a novel functionalization path through a supramolecular manner, where introduced functional groups always exist on the composite surface. Along the same lines, it is worth mentioning that an important aspect for the use of the chemically modified CUR is to functionalize the incorporated guest polymers by just wrapping them. In particular, the quantitative conversion of 6-OH groups to self-assembling groups allows the resultant composite to self-organize through the surface–surface interactions among the composites, where one-dimensional composites act as building blocks for creating the further hierarchical architecture.

4
Hierarchical Assemblies:
The One-Dimensional Composite
as a Building Block Toward Further Organization

Creation of highly ordered assemblies using functional polymers as building blocks, which would lead to novel chemical and physical properties depending on their assembling modes, is of great interest due to their potential applications as fundamental nanomaterials. The self-assembling system of small molecules have been well-established [195], whereas only a few attempts have been reported for the creation of such hierarchical architectures from polymers [196–198]. The difficulty in the polymeric system arises from how one can introduce self-assembling capabilities into a polymer backbone without losing its inherent functionality and from how one can assemble polymers through specific interpolymer interactions without the influence of the non-specific bundling nature. Furthermore, to tune physicochemical properties of polymer assemblies at nanoscale, individual polymers must be manipulated during their self-assembling processes. The unique hosting ability of β-1,3-glucans has several advantages for overcoming these difficulties in polymer manipulation: that is, (1) when chemically modified β-1,3-glucans are used as one-dimensional hosts, the exterior surface of the resultant nanocomposites can be utilized as an interaction site for the construction of supramolecular architectures and (2) the strong interpolymer interactions among guest polymers are perfectly suppressed by the wrapping effect of β-1,3-glucans, which insulates one piece of guest polymer to maintain its original functionality. As a preliminary example, two kinds of complementary semi-artificial CURs, i.e., CUR-N$^+$ and CUR-SO$_3^-$ were utilizing as a functional sheath for SWNTs, expecting that the mixture of these two composites in an appropriate ratio results in the creation of hierarchical SWNT architecture due to the electrostatic interaction [199]. It would be important to mention here that CUR-N$^+$ and CUR-SO$_3^-$ have similar wrapping capability for SWNTs, suggesting that SWNT/CUR-N$^+$ composite and SWNT/CUR-SO$_3^-$ composite may be used as "complementary" one-dimensional building blocks to create higher-order hierarchical self-assembled architectures (Fig. 38).

SWNT/CUR-N$^+$ and SWNT/CUR-SO$_3^-$ composite solutions containing the same concentration of SWNT were prepared according to the same procedure described in the previous section. The zeta-potential value of an aqueous solution containing SWNT/CUR-N$^+$ composite was estimated to be + 48.9 mV, whereas an aqueous solution containing SWNT/CUR-SO$_3^-$ composite showed – 49.5 mV. Once these two solutions were mixed in the same volume under very diluted conditions, the zeta-potential value of the resultant mixture showed – 0.53 mV without accompanying precipitate formation, indicating that the potential charges on these composites are almost neutralized to give a self-assembling composite through the electrostatic interaction.

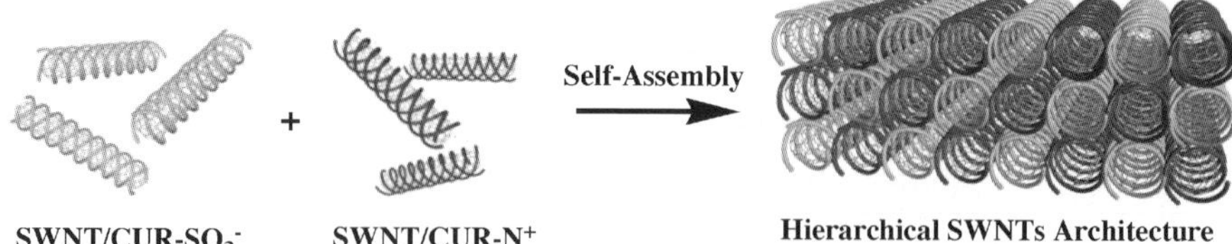

Fig. 38 Proposed concept for creating hierarchical SWNT architecture from one-dimensional building blocks through electrostatic interaction. It should be noted that all processes including wrapping of SWNTs by CUR and self-assembling of the resultant composites proceed in a supramolecular manner. Reprinted with permission from [199]

AFM images revealed that the resultant solution contains a well-developed sheet-like structure with micrometer-scale length, which is entirely different from the very-fine fibrous structures observed for individual SWNT/CUR-N$^+$ and SWNT/CUR-SO$_3^-$ composites (Fig. 39). These sheet-like structures show the characteristic Raman peaks at 262 and 1592 cm^{-1} being ascribed to SWNTs.

TEM is a powerful tool to study how SWNTs are arranged in the obtained sheet-like structure. In the TEM images shown in Fig. 40, it can be recognized that the sheet-like structure is composed of highly ordered fibrous assemblies. Furthermore, the electron diffraction pattern (inset in Fig. 40B) reveals that the fibrous assembly has some crystalline nature, suggesting that cationic and anionic composites are tightly packed through the electrostatic interaction. The periodicity of the dark layer is estimated to be ca. 2 nm, which is almost consistent with the diameter of the individual composite obtained by the AFM height profile. These results reasonably lead to the conclusion that a novel strategy toward the creation of "hierarchical" functional polymer architectures can be established by utilizing the complementary semi-artificial β-1,3-glucans as "building blocks".

The mixture of the oppositely charged small molecules tends to result in nonspecific irregular assemblies through electrostatic interactions, but when either a cationic or anionic polymer exists excessively, specific regular structures with well-controlled size and assembling number can be created. This concept may be applicable to the present polymer assembling system. When an aqueous SWNT/CUR-N$^+$ composite solution was mixed with an excess amount of an aqueous SWNT/CUR-SO$_3^-$ composite solution, adjusting the [SWNT/CUR-N$^+$]/[SWNT/CUR-SO$_3^-$] ratio to 1/5, bundled SWNTs architectures composed of highly ordered fibrous assemblies were obtained. The diameter of the bundle structure became larger, with increasing [SWNT/CUR-N$^+$]/[SWNT/CUR-SO$_3^-$] ratio from 1/5 to 1/3. These results indicate that the self-assembling hierarchical architecture is predictable and controllable by tuning the ratio of SWNT/CUR-N$^+$ composite and SWNT/CUR-SO$_3^-$ composite (Fig. 41).

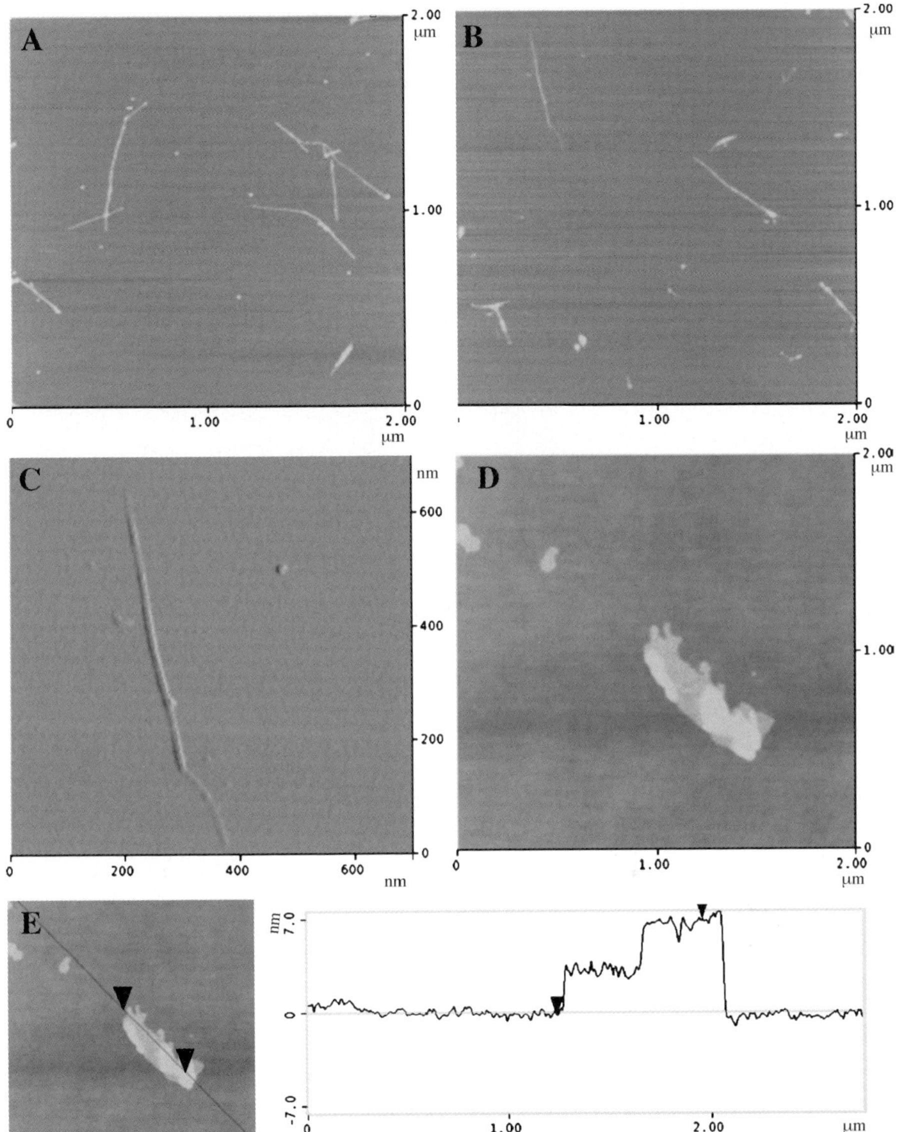

Fig. 39 AFM images of **A** SWNT/CUR-N$^+$ composite and **B** SWNT/CUR-SO$_3^-$ composite, respectively. **C** Magnified AFM image of (**B**). **D** AFM image of the sheet-like structure after mixing SWNT/CUR-N$^+$ composite and SWNT/CUR-SO$_3^-$ composite. **E** Height profile of the sheet-like structure: the AFM tip was scanned along the *black line*. In this AFM image, the thickness of each thin layer is estimated to be ca. 3.5 nm. Reprinted with permission from [199]

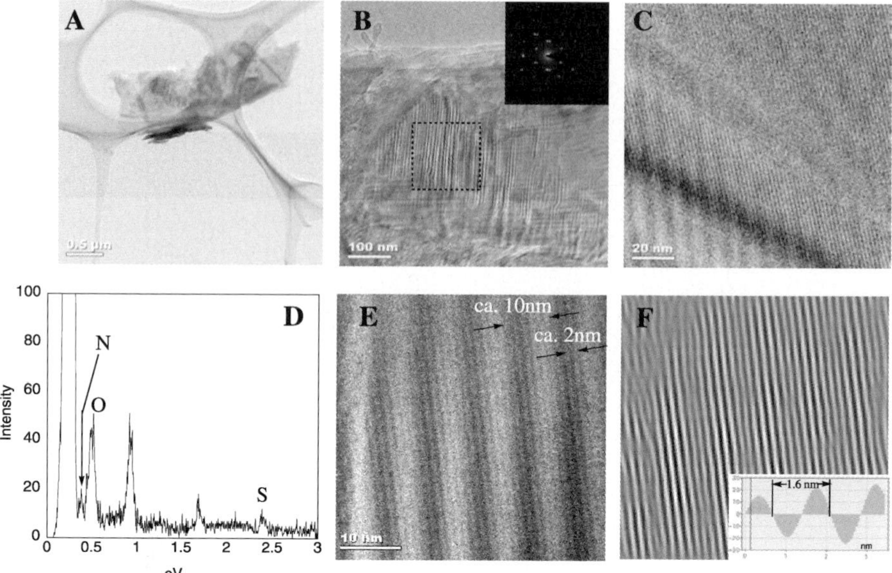

Fig. 40 TEM image of **A** Sheet-like structure (low magnification), **B**, **C** Magnified images of the thin layer (*inset*: electron diffraction pattern obtained from the sheet). **D** Elemental analysis of the sheet-like structure based on EDS. The spectrum was corrected from the *black-square* in (**B**). **E** Magnified TEM image of the sheet-like structure containing several thin layers. **F** Fourier translation image of (**D**) and extracted periodical patterns. Reprinted with permission from [199]

5
Conclusion and Outlook

Most polymer–polymer or polymer–molecule interactions, except those occurring in biological systems, have been considered to take place in a random fashion and to produce morphologically uninteresting polymer aggregates. In contrast, β-1,3-glucans can interact with polymer or molecular guests in a specific fashion and construct well-regulated one-dimensional superstructures: in the present system, we can predict how β-1,3-glucans wrap these guests. Furthermore, the wrapping occurs in an induced-fit manner, so that various functional nanocomposites can be created, reflecting the inherent functionalities of the entrapped guest material. These unique features of β-1,3-glucans mostly stem from the strong helix-forming nature and reversible interconversion between the single-stranded random coil and triple-stranded helix. It should be emphasized that the resultant composite can be applied to biomaterials due to the inherent bio-compatibility of β-1,3-glucans.

The clear wrapping mode allows us to utilize the composite as a one-dimensional building block for further molecular recognition events occurring on the composite surface: the selective modification of β-1,3-glucans

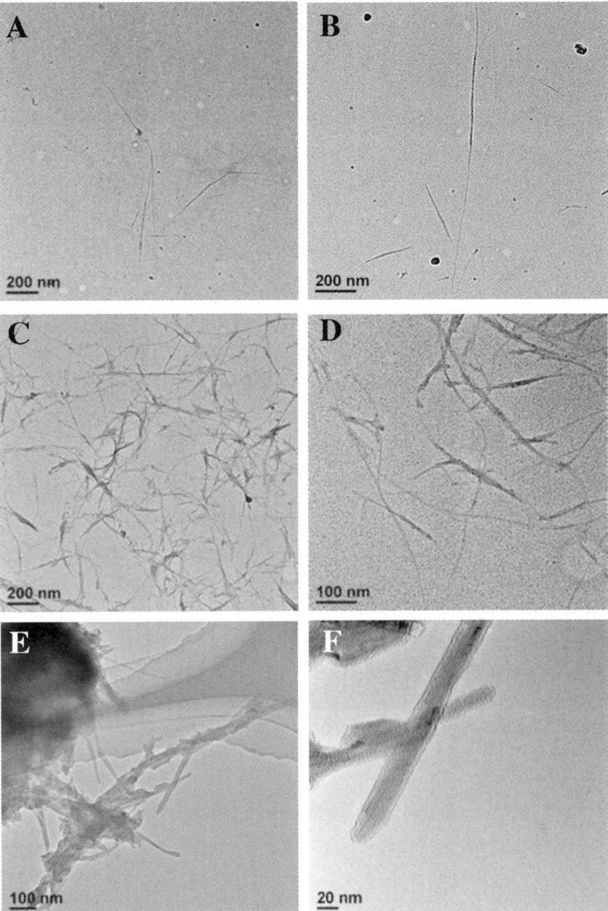

Fig. 41 TEM images of **A** SWNT/CUR-N$^+$ composite and **B** SWNT/CUR-SO$_3^-$ composite. **C** Fibrous bundle structure containing a limited number of SWNTs ([SWNT/CUR-N$^+$]/[SWNT/CUR-SO$_3^-$]=1/5). **D** Magnified TEM image of (**C**). **E** Larger bundle structure ([SWNT/CUR-N$^+$]/[SWNT/CUR-SO$_3^-$]=1/3). **F** Magnified TEM image of (**E**) (*inset*: extracted periodical patterns obtained along the *red line*). Reprinted with permission from [199]

endows the composite with molecular recognition and self-organization abilities. We have demonstrated that the wrapping of the chemically modified SPG provides a novel strategy to create functional polymer composites with various molecular recognition tags in a supramolecular manner. Especially, the quantitative conversion of 6-OH groups of CUR to self-assembling groups allows the resultant composite to self-organize through the specific surface–surface interactions among the composites, where the composite acts as one-dimensional building blocks for creating the further hierarchical architecture.

Considering the serious difficulties involved in the creation of hierarchical architectures from synthetic polymers, the present system can open up new paths to accelerate development of the polymer assembly systems and can extend the frontier of polysaccharide-based functional nanomaterials.

References

1. Rao VSR, Qasba PK, Balaji PV, Chandrasekaran R (1998) Conformation of Carbohydrates. Harwood Academic Publishers, The Netherlands
2. Takahashi Y, Kumano T, Nishikawa S (2004) Macromolecules 37:6827
3. Rappenecker G, Zugenmaier P (1981) Carbohydr Res 89:11
4. Wu HCH, Sarko A (1978) Carbohydr Res 61:7
5. Frampton MJ, Anderson HL (2007) Angew Chem Int Edit 46:1028
6. Immel S, Lichtenthaler FW (2000) Starch/Stärke 52:1
7. Hinrichs W, Büttner G, Steifa M, Betzel C, Zabel V, Pfannemüller B, Saenger W (1987) Science 238:205
8. French A, Zobel HF (1967) Biopolymers 5:457
9. Hellert W, Chanzy H (1994) J Biol Macromol 16:207
10. Rundle RE, Edwards FC (1943) J Am Chem Soc 65:2200
11. Kim OK, Choi LS (1994) Langmuir 10:2842
12. Choi LS, Kim OK (1998) Macromolecules 31:9406
13. Sanji T, Kato N, Tanaka M (2005) Chem Lett 34:1144
14. Sanji T, Kato N, Kato M, Tanaka M (2005) Angew Chem Int Edit 44:7301
15. Sanji T, Kato N, Tanaka M (2006) Org Lett 8:235
16. Kida T, Minabe T, Okabe S, Akashi M (2007) Chem Commun p 1559
17. Kadokawa J, Kaneko Y, Yagaya H, Chiba K (2001) Chem Commun p 459
18. Star A, Steuerman DW, Heath JR, Stoddart JF (2002) Angew Chem Int Edit 41:2508
19. Kim OK, Je J, Baldwin JW, Kooi S, Phehrsson PE, Buckley LJ (2003) J Am Chem Soc 125:4426
20. Atwood JL, Davies JED, MacNichol DD, Vögtle F (1996) Comprehensive Supramolecular Chemistry, vol 3. Pergamon, Oxford
21. Harada A, Li J, Kamachi M (1992) Nature 356:325
22. Harada A, Li J, Kamachi M (1994) Nature 370:126
23. Wenz G, Keller B (1992) Angew Chem Int Edit 31:197
24. Rekharsky MV, Inoue Y (1998) Chem Rev 98:1875
25. Li G, McGown LB (1994) Science 264:249
26. Pistolis G, Malliaris A (1996) J Phys Chem 100:15562
27. Michels JJ, O'Connell MJ, Taylor PN, Wilson JS, Cacialli F, Anderson HL (2003) Chem Eur J 9:6167
28. Gattuso G, Menzer S, Nepogodiev SA, Stoddart JF, Williams DJ (1997) Angew Chem Int Edit 36:1451
29. Harada A (1997) Adv Polym Sci 133:141
30. Harada A, Li J, Kamachi M (1993) Nature 364:516
31. Harada A (2001) Acc Chem Res 34:456
32. Wenz G, Han BH, Müller A (2006) Chem Rev 106:782
33. Raymo FM, Stoddart JF (1999) Chem Rev 99:1643
34. Nepogodiev SA, Stoddart JF (1998) Chem Rev 98:1959

35. van den Boogaard M, Bonnet G, van't Hof P, Wang Y, Brochon C, van Hutten P, Hadziioannou G (2004) Chem Mater 16:4383
36. Shen X, Belletete M, Durocher G (1998) Chem Phys Lett 298:201
37. Lagrost C, Ching KIC, Lacroix JC, Aeiyach S, Jouini M, Lacaze PC, Tanguy J (1999) J Mater Chem 9:2351
38. Takashima Y, Oizumi Y, Sakamoto K, Miyauchi M, Kamitori S, Harada A (2004) Macromolecules 37:3962
39. Hapiot P, Lagrost C, Aeiyach S, Jouini M, Lacroix JC (2002) J Phys Chem B 106:3622
40. Lagrost C, Tanguy J, Aeiyach S, Lacroix JC, Jouini M, Chane-Ching KI, Lacaze PC (1999) J Electroanal Chem 476:1
41. Yamaguchi I, Kashiwagi K, Yamamoto T (2004) Macromol Rapid Commun 25:1163
42. Belosludov RV, Sato H, Farajian AA, Mizuseki H, Ichinoseki K, Kawazoe Y (2004) Japan J Appl Phys 42:2492
43. do Nascimento GM, da Silva JEP, de Torresi SIC, Santos PS, Temperini MLA (2002) Mol Cryst Liq Cryst 374:53
44. Yoshida K, Shimomura T, Ito K, Hayakawa R (1999) Langmuir 15:910
45. Akai T, Abe T, Shimomura T, Ito K (2001) Japan J Appl Phys 40:L1327
46. Shimomura T, Akai T, Abe T, Ito K (2002) J Chem Phys 116:1753
47. Shimomura T, Yoshida K, Ito K, Hayakawa R (2000) Polym Adv Technol 11:837
48. Okumura H, Kawaguchi Y, Harada A (2001) Macromolecules 34:6338
49. Okumura H, Kawaguchi Y, Harada A (2002) Macromol Rapid Commun 23:781
50. Okumura H, Kawaguchi Y, Harada A (2003) Macromolecules 36:6422
51. Yanaki T, Norisuye T, Fujita H (1980) Macromolecules 13:1462
52. van der Valk P, Marchant R, Wessels JGH (1977) Exp Mycol 1:69
53. Yanaki T, Ito W, Tabata K, Kojima T, Norisuye T, Takano N, Fujita H (1983) Biophys Chem 17:337
54. Bohn JA, BeMiller JN (1995) Carbohydr Polym 28:3
55. Sato T, Norisuye T, Fujita H (1981) Carbohydr Res 95:195
56. Stokke BT, Elgsaeter A, Brant DA, Kitamura S (1991) Macromolecules 24:6349
57. Sato T, Norisuye T, Fujita H (1983) Macromolecules 16:185
58. McIntire TM, Brant D (1998) J Am Chem Soc 120:6909
59. Bluhm TL, Deslandes Y, Marchessault RH (1982) Carbohydr Res 100:117
60. Harada T, Misaki A, Saito H (1968) Arch Biochem Biophys 124:292
61. Deslandes Y, Marchessault RH, Sarko A (1980) Macromolecules 13:1466
62. Frecer V, Rizzo R, Miertus S (2000) Biomacromolecules 1:91
63. Chuah CT, Sarko A, Deslandes Y, Marchessault RH (1983) Macromolecules 16:1375
64. Atkins EDT, Parker KD (1968) Nature 220:784
65. Marchessault RH, Deslandes Y, Ogawa K, Sundararajan PR (1977) Can J Chem 55:300
66. Norisuye T, Yanaki T, Fujita H (1980) J Polym Sci Polym Phys Edit 18:547
67. Kashiwagi Y, Norisuye T, Fujita H (1981) Macromolecules 14:1220
68. Sakurai K, Shinkai S (2000) J Am Chem Soc 122:4520
69. Sakurai K, Uezu K, Numata M, Hasegawa T, Li C, Kaneko K, Shinkai S (2005) Chem Commun p 4383
70. Iijima S (1991) Nature 354:56
71. Iijima S, Ichihashi T (1993) Nature 363:603
72. Bethune DS, Kiang CH, de Vries MS, Gorman G, Savoy R, Savoy J (1993) Nature 363:605
73. Zheng M, Jagota A, Semke ED, Diner BA, Mclean RS, Lustig SR, Tassi RER (2003) Nat Mater 2:338

74. Zheng M, Jagota A, Strano MS, Santos AP, Barone P, Chou SG, Diner BA, Dressel-haus MS, Mclean RS, Onoa GB, Samsonidze GG, Semke ED, Usrey M, Walls DJ (2003) Science 302:1545
75. Nakashima N, Okuzono S, Murakami H, Nakai T, Yoshikawa K (2003) Chem Lett 32:456
76. Takahashi T, Tsunoda K, Yajima H, Ishii T (2002) Chem Lett 31:690
77. Pantarotto D, Partidos CD, Graff R, Hoebeke J, Briand JP, Prato M, Bianco A (2003) J Am Chem Soc 125:6160
78. Dieckmann GR, Dalton AB, Johnson PA, Razal J, Chen J, Giordano GM, Munoz E, Musselman IH, Baughaman RH, Draper RK (2003) J Am Chem Soc 125:1770
79. Zorbas V, Ortiz-Acevedo A, Dalton AB, Yoshida MM, Dieckmann GR, Draper RK, Baughman RH, Yacaman MJ, Musselman IM (2004) J Am Chem Soc 126:7222
80. Sano M, Kamino A, Okamura J, Shinkai S (2001) Science 293:1299
81. Ziegler KJ, Gu Z, Peng H, Flor EL, Hauge RH, Smalley RE (2005) J Am Chem Soc 127:1541
82. Liu J, Rinzler AG, Dai HJ, Hafner JH, Bradley RK, Boul PJ, Lu A, Iverson T, Shelimov K, Huffman CB, Rodriguez-Macias F, Shon YS, Lee TR, Colbert DT, Smalley RE (1998) Science 280:1253
83. Zhang J, Zou H, Qing Q, Yang Y, Li Q, Liu Z, Guo X, Du Z (2003) J Phys Chem B 107:3712
84. Numata M, Asai M, Kaneko K, Hasegawa T, Fujita N, Kitada Y, Sakurai K, Shinkai S (2004) Chem Lett 33:232
85. Kimura T, Koumoto K, Sakurai K, Shinkai S (2000) Chem Lett 29:1242
86. Numata M, Asai M, Kaneko K, Bae AH, Hasegawa T, Sakurai K, Shinkai S (2005) J Am Chem Soc 127:5875
87. Coleman JN, Ferreira MS (2004) Appl Phys Lett 84:798
88. Midgley PA, Thomas JM, Weyland M, Johnson BFG (2001) Chem Commun p 907
89. Madalia O, Weber I, Frangakis AS, Gerisch G, Baumeister W (2002) Science 298:1209
90. Thomas JM, Midgley PA (2004) Chem Commun p 1253
91. Kermer JR, Mastronarde DN, McIntosh JR (1996) J Struct Biol 116:71
92. Frank J (1996) Three-dimensional Electron Microscopy of Macromolecular Assemblies. Academic Press, San Diego
93. Jinnai H, Nishikawa Y, Spantak RJ, Smith SD, Agard DA, Hashimoto T (2000) Phys Rev Lett 84:518
94. MacDiamid AG, Chiang JC, Halpern M, Huang WS, Mu SL, Somasiri NLD, Wu W, Yaniger SI (1985) Mol Cryst Liq Cryst 121:173
95. MacDiamid AG, Chiang JC, Richter AF, Epstein AJ (1987) Synth Met 18:285
96. Lu FL, Wudl F, Nowak M, Heeger AJ (1986) J Am Chem Soc 108:8311
97. Numata M, Hasegawa T, Fujisawa T, Sakurai K, Shinkai S (2004) Org Lett 6:4447
98. Buey J, Swager TM (2000) Angew Chem Int Edit 39:608
99. Sauvage JP, Kern JM, Bidan G, Divisia-Blohorn B, Vidal PL (2002) New J Chem 26:1287
100. Martin RE, Diederich F (1999) Angew Chem Int Edit 38:1350
101. Hecht S, Frechet JMJ (2001) Angew Chem Int Edit 40:74
102. Sato T, Jiang DL, Aida T (1999) J Am Chem Soc 121:10658
103. Li C, Numata M, Bae AH, Sakurai K, Shinkai S (2005) J Am Chem Soc 127:4548
104. Langeveld-Voss BMW, Janssen RAJ, Christiaans MPT, Meskers SCJ, Dekkers HPJM, Meijer EM (1996) J Am Chem Soc 118:4908
105. Goto H, Okamoto Y, Yashima E (2002) Chem Eur J 8:4027
106. Kim J, McQuade DT, McHugh SK, Swager TM (2000) Angew Chem Int Edit 39:3868

107. Tan C, Atas E, Müller JG, Pinto MR, Kleiman VD, Schanze KS (2004) J Am Chem Soc 126:13685
108. Zahn S, Swager TM (2002) Angew Chem Int Edit 41:4226
109. DiCesare N, Pinto MR, Schanze KS, Lakowica JR (2002) Langmuir 18:7785
110. Kumaraswamy S, Bergstedt T, Shi X, Rininsland F, Kushon S, Xia W, Ley K, Achyuthan K, McBranch D, Whitten D (2004) Proc Natl Acad Sci USA 101:7511
111. Pinto MR, Schanze KS (2004) Proc Natl Acad Sci USA 101:7505
112. Fan C, Plaxco KW, Heeger AJ (2002) J Am Chem Soc 124:5642
113. Wosnick JH, Mello CM, Swager TM (2005) J Am Chem Soc 127:3400
114. Numata M, Fujisawa T, Li C, Haraguchi S, Ikeda M, Sakura K, Shinkai S (2007) Supramol Chem 19:107
115. Ichino Y, Minami N, Yatabe T, Obata K, Kira M (2001) J Phys Chem 105:4111
116. Nakashima H, Fujiki M, Koe JR, Motonaga M (2001) J Am Chem Soc 123:1963
117. Peng W, Motonaga M, Koe JR (2004) J Am Chem Soc 126:13822
118. Haraguchi S, Hasegawa T, Numata M, Fujiki M, Uezu K, Sakurai K, Shinkai S (2005) Org Lett 7:5605
119. Fujiki M (1994) J Am Chem Soc 116:6017
120. Würthner F, Yao S, Beginn U (2003) Angew Chem Int Edit 42:3247
121. Parins LJ, Thalacker C, Würthner F, Timmerman P, Reinhoudt DN (2001) Proc Natl Acad Sci USA 98:10042
122. Jones RM, Lu L, Helgeson R, Bergstedt TS, McBranch DW, Whitten DG (2001) Proc Natl Acad Sci USA 98:14769
123. Yao S, Beginn U, Gress T, Lysetska M, Würthner F (2004) J Am Chem Soc 126:8336
124. Morikawa M, Yoshihara M, Endo T, Kimizuka N (2005) J Am Chem Soc 127:1358
125. Hannah KC, Armitage BA (2004) Acc Chem Res 37:845
126. Heuer WB, Kim OK (1998) Chem Commun p 2649
127. Clays K, Olbrechts G, Munters T, Persoons A, Kim OK, Choi LS (1998) Chem Phys Lett 293:337
128. Kim OK, Je J, Jernigan G, Buckley L, Whitte D (2006) J Am Chem Soc 128:510
129. Kashida H, Asanuma H, Komiyama M (2006) Chem Commun p 2768
130. Kashida H, Asanuma H, Komiyama M (2004) Angew Chem Int Edit 43:6522
131. Kool ET (2002) Acc Chem Res 35:936
132. Onouchi H, Miyagawa T, Morino K, Yashima E (2006) Angew Chem Int Edit 45:2381
133. Li C, Numata M, Takeuchi M, Shinkai S (2005) Angew Chem Int Edit 44:6371
134. Endo M, Wang H, Fujitsuka M, Majima T (2006) Chem Eur J 12:3735
135. Koti ASR, Periasamy N (2003) Chem Mater 15:369
136. Wang M, Silva GL, Armitage BA (2000) J Am Chem Soc 122:9977
137. Renikuntala BR, Armitage BA (2005) Langmuir 21:5362
138. Garoff RA, Litzinger EA, Connor RE, Fishman I, Armitage BA (2002) Langmuir 18:6330
139. Aoki K, Nakagawa M, Ichimura K (2000) J Am Chem Soc 122:10997
140. Nakagawa M, Ishii D, Aoki K, Seki T, Iyoda T (2005) Adv Mater 17:200
141. Numata M, Tamesue S, Fujisawa T, Haraguchi S, Hasegawa T, Bae AH, Li C, Sakurai K, Shinkai S (2006) Org Lett 8:5533
142. Hasegawa T, Fujisawa T, Numata M, Li C, Bae AH, Haraguchi S, Sakurai K, Shinkai S (2005) Chem Lett 34:1118
143. Morigaki K, Baumgart T, Jonas U, Offenhausser A, Knoll W (2002) Langmuir 18:4082
144. Niwa M, Shibahara S, Higashi N (2000) J Mater Chem 10:2647
145. Menzel H, Mowery MD, Cai M, Evans CE (1999) Adv Mater 11:131
146. Tajima K, Aida T (2000) Chem Commun p 2399

147. Ying JY, Mehnert CP, Wong MS (1999) Angew Chem Int Edit 38:56
148. Uemura T, Horiike S, Kitagawa S (2006) Chem Asian J 1:36
149. Hasegawa T, Haraguchi S, Numata M, Fujisawa T, Li C, Kaneko K, Sakurai K, Shinkai S (2005) Chem Lett 34:40
150. Hasegawa T, Haraguchi S, Numata M, Li C, Bae AH, Fujisawa T, Kaneko K, Sakurai K, Shinkai S (2005) Org Biomol Chem 3:4321
151. Wenz G, Müller MA, Schmidt M, Wegner G (1984) Macromolecules 17:837
152. Shirai E, Urai Y, Itoh K (1998) J Phys Chem B 102:3765
153. Soler-lllia GJ de AA, Sanchez C, Lebeau B, Patarin J (2002) Chem Rev 102:4093
154. Davis SA, Breulmann M, Rhodes KH, Zhang B, Mann S (2001) Chem Mater 13:3218
155. van Bommel KJC, Friggeri A, Shinkai S (2003) Angew Chem Int Edit 42:980
156. Huang J, Kunitake T (2003), J Am Chem Soc 125:11834
157. Ichinose I, Hashimoto Y, Kunitake T (2004) Chem Lett 33:656
158. Antonietti M, Breulmann M, Göltner CG, Cölfen H, Wong KK, Walsh D, Mann S (1998) Chem Eur J 4:2493
159. Qi L, Cölfen H, Antonietti M, Li M, Hopwood JD, Ashley AJ, Mann S (2001) Chem Eur J 7:3526
160. Numata M, Li C, Bae AH, Kaneko K, Sakurai K, Shinkai S (2005) Chem Commun p 4655
161. Groenendaal L, Jonas F, Freitag D, Pielartzik H, Reynolds JR (2000) Adv Mater 12:481
162. Li C, Imae T (2004) Macromolecules 37:2411
163. Li C, Hatano T, Takeuchi M, Shinkai S (2004) Tetrahedron 60:8037
164. Pringsheim E, Terpetschnig E, Piletsky SA, Wolfbeis OS (1999) Adv Mater 11:865
165. Kim BH, Kim MS, Park KT, Lee JK, Park DH, Joo J, Yu SG, Lee SH (2003) Appl Phys Lett 83:539
166. Hulvat JF, Stupp SI (2004) Adv Mater 16:589
167. Goto H, Akagi K (2004) Macromol Rapid Commun 25:1482
168. Han MG, Armes SP (2003) Langmuir 19:4523
169. Han MG, Foulger SH (2004) Adv Mater 16:231
170. Li C, Numata M, Hasegawa T, Fujisawa T, Haracuchi S, Sakurai K, Shinkai S (2005) Chem Lett 34:1532
171. Tour JM (1996) Chem Rev 96:537
172. Roncali J (1997) Chem Rev 97:173
173. McQuade DT, Pullen AE, Swager TM (2000) Chem Rev 100:2537
174. Katz E, Willner I (2004) Angew Chem Int Edit 43:6042
175. Devries GA, Brunnbauer M, Hu Y, Jackson AM, Long B, Neltner BT, Uzun O, Wunsch BH, Stellacci F (2007) Science 315:358
176. Bae AH, Numata M, Hasegawa T, Li C, Kaneko K, Sakurai K, Shinkai S (2005) Angew Chem Int Edit 44:2030
177. Bae AH, Numata M, Yamada S, Shinkai S (2007) New J Chem 31:618
178. Demleitner S, Kraus J, Franz G (1992) Carbohydr Res 226:239
179. Gao Y, Katsuraya K, Kaneko Y, Mimura T, Nakashima H, Uryu T (1999) Macromolecules 32:8319
180. Katsuraya K, Nakashima H, Yamamoto N, Uryu T (1999) Carbohydr Res 315:234
181. Yoshida T, Yasuda Y, Mimura T, Kaneko Y, Nakashima H, Yamamoto N, Uryu T (1995) Carbohydr Res 276:425
182. Katsuraya K, Ikushima N, Takahashi N, Shoji T, Nakashima H, Yamamoto N, Yoshida T, Uryu T (1994) Carbohydr Res 260:51
183. Na K, Park KH, Kim SW, Bae YH (2000) J Control Release 69:225
184. Koumoto K, Umeda M, Numata M, Matsumoto T, Sakurai K, Kunitake T, Shinkai S (2004) Carbohydr Res 339:161

185. Hasegawa T, Umeda M, Numata M, Fujisawa T, Haraguchi S, Sakurai K, Shinkai S (2006) Chem Lett 35:82
186. Hasegawa T, Umeda M, Numata M, Li C, Bae AH, Fujisawa T, Haraguchi S, Sakurai K, Shinkai S (2006) Carbohydr Res 341:35
187. Numata M, Matsumoto T, Umeda M, Koumoto K, Sakurai K, Shinkai S (2003) Bioorg Chem 163:171
188. Hasagawa T, Umeda M, Matsumoto T, Numata M, Mizu M, Koumoto K, Sakurai K, Shinkai S (2004) Chem Commun p 382
189. Hasegawa T, Fujisawa T, Numata M, Matsumoto T, Umeda M, Karinaga R, Mizu M, Koumoto K, Kimura T, Okumura S, Sakurai K, Shinkai S (2004) Org Biomol Chem 2:3091
190. Hasegawa T, Fujisawa T, Numata M, Umeda M, MatsumotoT, Kimura T, Okumura S, Sakurai K, Shinkai S (2004) Chem Commun p 2150
191. Hasegawa T, Fujisawa T, Haraguchi S, Numata M, Karinaga R, Kimura T, Okumura S, Sakurai K, Shinkai S (2005) Bioorg Med Chem Lett 15:327
192. Ikeda M, Hasegawa T, Numata M, Sugikawa K, Sakurai K, Fujiki M, Shinkai S (2007) J Am Chem Soc 129:3979
193. Ikeda M, Minari J, Shimada N, Numata M, Sakurai K, Shinkai S (2007) Org Biomol Chem 5:2219
194. Ikeda M, Haraguchi S, Numata M, Shinkai S (2007) Chem Asian J 2:1290
195. Lehn JM (1995) Supramoleculer Chemistry. VCH, New York
196. Kubo Y, Kitada Y, Wakabayashi R, Kishida T, Ayabe M, Kaneko K, Takeuchi M, Shinkai S (2006) Angew Chem Int Edit 45:1548
197. Ikeda M, Furusho Y, Okoshi K, Tanahara S, Maeda K, Nishino S, Mori T, Yashima E (2006) Angew Chem Int Edit 45:6491
198. Ewert KK, Evans HM, Zidovska A, Bouxsein NF, Ahmad A, Safinya CR (2006) J Am Chem Soc 128:3998
199. Numata M, Sugikawa K, Kaneko K, Shinkai S (2008) Chem Eur J 14:2398

Adv Polym Sci (2008) 220: 123–187
DOI 10.1007/12_2008_142
© Springer-Verlag Berlin Heidelberg
Published online: 11 April 2008

Supramolecular Organization of Polymeric Materials in Nanoporous Hard Templates

Martin Steinhart

Max Planck Institute of Microstructure Physics, Weinberg 2, 06120 Halle, Germany
steinhart@mpi-halle.de

Abstract A broad range of polymeric materials can be formed into nanotubes by means of nanoporous hard templates containing arrays of aligned, cylindrical nanopores. Functional hybrid membranes consisting of the nanoporous matrix and the nanotubes as well as released arrays of aligned nanotubes are thus accessible. The mechanical, chemical, optical, and electronic properties of the nanotubes as well as their specific surface are largely determined by the supramolecular organization of the material they consist of, and only the rational design of their internal morphology will pave the pay for their use as functional device components. Herein, recent efforts to tailor the mesoscopic structure of nanotubes by controlling the way precursors and target materials are deposited into the nanopores are reviewed. Moreover, specific attention is directed to structure formation processes such as crystallization, phase separation and mesophase formation under the influence of the two-dimensional confinement imposed by the pore geometry and the interfacial interactions with the pore walls. Nanoporous hard templates are particularly suitable for the rational generation of mesocopic fine structures in nanofibers because equilibrium and non-equilibrium states as well as unprecedented confinement-induced morphologies with new and exciting properties can be realized.

Keywords Nanotubes · Porous templates · Self-assembly · Wetting

Abbreviations

AAO	Anodic aluminum oxide
BCP	Block copolymer
DMF	Dimethylformamide
D_p	Pore diameter
DSC	Differential scanning calorimetry
IR	Infrared
HA	Hard anodization
HRTEM	High-resolution transmission electron microscopy
L_0	Bulk period of a BCP
LC	Liquid crystal
MA	Mild anodization
M_W	Weight-average molecular weight
NMR	Nuclear magnetic resonance
Pluronic F-127	Ethyleneoxide$_{106}$-propyleneoxide$_{70}$-ethyleneoxide$_{106}$
PAN	Poly(acrylonitrile)
Pd	Palladium
PC	Polycarbonate
PDMS	Poly(dimethylsiloxane)
PE	Poly(ethylene)
PEO	Poly(ethylene oxide)
PL	Photoluminescence
PLA	Poly(lactide)
PMMA	Poly(methyl methacrylate)
PPO	Poly(propylene oxide)
PS	Poly(styrene)
PS-*b*-PBD	Poly(styrene-*block*-butadiene)
PS-*b*-PMMA	Poly(styrene-*block*-methyl methacrylate)
Pt	Platinum
PVDF	Poly(vinylidene difluoride)
SAED	Selected-area electron diffraction
SAXS	Small-angle X-ray scattering
SEM	Scanning electron microscopy
TEM	Transmission electron microscopy
TEOS	Tetraethoxysilane
T_c	Crystallization temperature
THF	Tetrahydrofuran
T_M	Melting temperature
T_p	Pore depth
WAXS	Wide angle X-ray scattering

1
Introduction

The range of materials that can be formed into nanotubes has been significantly extended during the past decade. However, it is still a challenge to tailor the internal fine structure of the nanotube walls by controlling structure formation processes such as crystallization, mesophase formation and phase separation. The self-organization of the material the nanotube walls consist of on the supramolecular scale largely determines the chemical as well as the physical properties of the nanotubes and therefore their potential for real-life applications. For example, crystal size and crystal orientation in the walls of nanotubes consisting of semicrystalline polymers will largely influence their optical, electronic, mechanical and ferroelectric properties. As discussed below, mesoporous nanofibers that contain arrays of aligned cylindrical pores or hollow spaces with non-conventional geometries, such as helical pores, can be produced by the self-assembly of molecular block copolymer (BCP) soft templates under varying degrees of geometric confinement. The emerging interest in such "complex" nanotubes suggests that the traditional distinction between "nanotubes" and "nanorods" is somewhat arbitrary, if not obsolete. On the contrary, the design of the mesoscopic, supramolecular constitution of the one-dimensional nanostructures appears to be by far more crucial for their properties than the presence or absence of a hollow space in their interior. It is therefore highly desirable to control the formation of the internal morphology during the preparation of the nanotubes as far as possible.

Two fundamentally different strategies, both of which are associated with certain advantages and drawbacks, allow fabricating nanotubes of virtually any functional material. The first strategy, which is addressed in other parts of this volume, involves the direct self-assembly of molecular and supramolecular building blocks, or the use of soft templates that direct the formation of tubular structures from specific target materials or precursors thereof. The experimental configuration is simple because all components required for the synthesis are contained in one and the same solution, and self-assembly resulting in the formation of nanotubes often takes place under mild conditions. The nanotubes can be produced in large quantities, and their separation as well as their purification is possible with common methods such as filtration and centrifugation. However, the target materials the nanotubes consist of (or the corresponding precursors) must exhibit an intrinsic ability to self-assemble or they must interact with a structure-directing soft template in a very specific way. The range of materials that show these properties is limited. Moreover, it is difficult to arrange the nanotubes thus obtained into ordered arrays, as required for their integration into functional device architectures.

A complementary approach to the fabrication of nanotubes involves the use of hard templates as tools. Hard templates are either nanofibers or porous host materials. In the former case, the nanofibers are at first coated with the wall material of the tubes or a corresponding precursor. Subsequently, the template fiber, that is, the core of the hybrid fiber thus obtained, is selectively removed so that a shell of the material initially deposited onto the template nanofiber is conserved. Template fibers can, for example, be produced in high quantities and with high aspect ratios (ratio of length and diameter) by electrospinning. Consequently, their use as templates makes non-woven fabrics consisting of nanotubes with high aspect ratios accessible, which are highly interesting materials for packaging, thermal insulation, storage, separation and high-performance filters [1].

However, if arrays of aligned nanotubes with precisely adjustable aspect ratios are required, nanoporous host materials exhibiting arrays of aligned nanochannels are the template of choice. Several kinds of nanoporous membranes [2] have been used as shape-defining molds, predominantly track-etch membranes and nanoporous anodic aluminum oxide (AAO), which can therefore be referred to as hard templates. Their use yields tubular structures from a broad range of polymeric materials, whose arrangement is determined by that of the pores in the hard template. The material the nanotubes consist of can directly be deposited onto the pore walls. It is also possible to infiltrate precursors or monomers into the nanoporous hard templates and to convert them into the target materials. A comprehensive body of literature, including many excellent review articles, deals with this topic [3–5]. The preparation of nanotubes inside the pores of nanoporous hard templates, which was pioneered by Martin and co-workers [6–14], automatically yields porous hybrid membranes whose channels are functionalized with the nanotubes in their interior. Examples for this are DNA-functionalized nanotube membranes with single-base mismatch selectivity [15] or membranes for enantioselective separations [16]. A problem associated with the fabrication of nanotubes by means of hard templates is the required availability of the latter. The release of the nanotubes is commonly achieved by a wet-chemical etching step destroying the hard template, which is a drawback for the up-scaling of template-based approaches to the fabrication of nanotubes. If they are attached to a support, they form more or less ordered arrays. Such arrays are of interest since they may exhibit specific wetting and adhesive properties [17, 18]. Recently reported approaches to the mechanical extraction of fiber arrays from porous templates [19, 20] need to be optimized and require that the nanotubes are tightly connected with an underlying substrate.

Despite these still-challenging issues, the fabrication of nanotubes using nanoporous hard templates is associated with several advantages beyond the possibility to align them. Readily available porous hosts such as self-ordered porous AAO have a narrow pore diameter (D_p) distribution and pores with D_p-values ranging from about 20 nm up to a few 100 nm. The pore depths (T_p) can be adjusted to values between about 1 μm and several 100 μm. Therefore, it is easily possible to tailor the diameter and the aspect ratio of the nanotubes. However, the most important advantage is the possibility to control mesoscopic structure formation processes inside the pores. There are relatively few limitations regarding the materials that can be formed into nanotubes via hard templates. Mixtures, sols, semicrystalline and liquid-crystalline polymers, other thermoplastics, as well as BCPs, are eligible for this approach so that the mesoscopic structure formation processes these materials undergo, such as phase separation, crystallization and mesophase formation, can be exploited to rationally generate specific, self-assembled supramolecular architectures in the tube walls. The degree of geometric confinement can be adjusted by the D_p-value. The chemical properties of the rigid pore walls of the hard template can by modified too, for example, by silanization or atomic layer deposition [21, 22]. The high surface-to-volume ratio of the nanoporous hard templates makes it possible to control self-assembly processes by adjusting the properties of the pore walls. This is particularly the case for self-assembly processes based on phase separation, as discussed below. Moreover, the walls of the template pores can be functionalized in such a way that they are electrically charged, a prerequisite for the fabrication of nanotubes by layer-by-layer deposition of polyelectrolytes. As common hard templates consist of inorganic materials such as alumina, they are stable at temperatures at which molten polymers are typically processed. There are no limitations regarding the temperature profile applied to the samples, that is, molten polymers can be crystallized either isothermally or non-isothermally. Hence, using hard templates in the synthesis of nanotubes is associated with various handles to tailor the internal fine structure of their walls.

The organization herein is as follows. Commonly used hard template systems will be described in Sect. 2. Section 3 deals with the infiltration of the target materials the nanotubes consist of or corresponding precursors into the pores. This step is far from being trivial. Even though the underlying physico-chemical phenomena are not completely understood, a qualitative overview of the infiltration mechanisms will be given. Crystallization and the formation of mesophases from liquid-crystalline molecular building blocks are important examples of self-organization processes of single-component materials that can be exploited to generate textured nanofibers (Sect. 4). As discussed below, not only the confinement but also the presence or absence of a bulk reservoir that is connected with the nanofibers influences the crystallization of polymers, which form lamella crystals and spherulitic superstructures in the bulk, inside the nanopores of hard templates. The texture in liquid-crystalline pore walls is the result of a delicate interplay of the anchoring to the pore walls and the growth kinetics. The exploitation of self-organization processes based on phase separation is discussed in Sect. 5. This includes the generation of fine structures by spinodal decomposition of mixtures as well as the formation of "complex" nanotubes and nanofibers with non-conventional morphologies by self-assembling BCPs inside the template pores. In this connection, we will also cover the synthesis of mesoporous nanofibers consisting of silica and amorphous carbon, as the underlying self-assembly processes are largely related to the BCP soft templates used as structure-directing agents. Section 6 deals with the preparation of nanotubes by layer-by-layer deposition of polyelectrolytes and other polymeric materials into nanoporous hard templates. This technique involves the successive deposition of different target materials onto the pore walls, taking advantage of specific interactions between the species to be deposited and the already deposited layers. Thus, nanotubes having walls with a well-defined multilayer structure are accessible. This methodology also allows fabricating nanotubes composed of complex macromolecules such as dendrimeric polyelectrolytes and the controlled incorporation of nanoparticles such as luminescent quantum dots into the nanotube walls.

2
Hard Templates

A prerequisite for the template-based production of one-dimensional nanostructures and the rational exploitation of self-assembly processes in the two-dimensional confinement imposed by the pore geometry is the availability of suitable hard templates. In this section, a brief overview of commonly used hard templates, of their advantages and of their disadvantages will be given. However, it is beyond the scope of this contribution to exhaustively review the fabrication and properties of these porous materials. Commonly, two types of hard templates, both of which can contain arrays of aligned cylindrical channels as well as more complex pore shapes, are employed in the production of nanotubes and nanorods: track-etch polymer membranes and nanoporous AAO. Track-etch membranes [23, 24] (Fig. 1) are produced by irradiating polymeric films with a thickness ranging from a few microns to a few tens of microns with ion beams, thus producing latent tracks penetrating through the bombarded films. In a second step, pores are generated at the positions of the latent tracks by wet-chemical etching. Pore size, shape and density can be varied in a controllable manner by the proper selection of the conditions under which irradiation and post-treatment procedures are carried out. Pores with D_p-values ranging from 10 nm to the micron range are

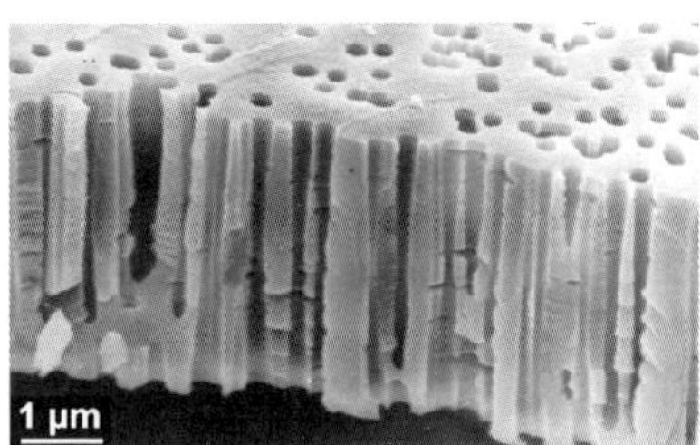

Fig. 1 Example of a polymeric track-etch membrane. Reproduced from [23]. © (2001) Elsevier

obtained, whereas the pore density can be adjusted to any value between 1 to 10^{10} pores per centimeter. Moreover, D_p-value and pore density can be adjusted independently. The most common polymers track-etch membranes consist of are polycarbonate and polyethylene terephthalate. The pore walls are commonly hydrophilized by plasma treatment or by adsorbing or grafting hydrophilic polymers, such as polyvinyl pyrrolidone, onto the pore walls. The limitations associated with track-etch membranes are their limited stability at elevated temperatures and their poor resistance against organic solvents, which poses problems for many of the self-assembly processes discussed below. The arrangement of the pores is random, that is, track-etch membranes do not exhibit long-range order. Moreover, because of their poor rigidity and their lack of chemical resistance to organic solvents, it is difficult to remove residual material from the surface of track-etch membranes after their infiltration, a process step that is crucial to the template-based fabrication of nanotubes and nanorods. Nevertheless, because of their commercial availability and versatility, track-etch membranes are being routinely used for the production of one-dimensional nanostructures. However, it was found that the pronounced roughness of the pore walls in track-etch membranes revealed by SEM and adsorption experiments prevents uniform orientation of anisotropic species infiltrated into the pores [24].

The second common hard template is nanoporous AAO with hydroxyl-terminated pore walls produced by the electrochemical anodization of aluminum substrates. From a practical viewpoint AAO has several advantages. It is stable at temperatures at which soft matter is commonly processed and resistant against organic solvents but can be selectively etched with aqueous acids and bases to release nanofibers fabricated inside its pores. The basis for the production of AAO is the well-known formation of porous oxide layers, whose thickness increases linearly with the anodization time, on bulk aluminum. To this end, electrolyte solutions are used that partially dissolve the freshly formed alumina [25–27]. At first, a homogeneous barrier oxide layer is generated on the aluminum substrate. Field-enhanced dissolution of the oxide occurs at fluctuations in the oxide layer, which leads in turn to the formation of pores. As the pore density increases, an array of pores characterized by a mean interpore distance develops. A stationary state in which pore growth is characterized by an interplay of field-enhanced dissolution of alumina at the pore bottoms and the formation of new alumina leads to stable pore growth [28, 29]. In AAO prepared under so-called mild anodization (MA) conditions, the amorphous pore walls consist of an outer layer containing water, electrolyte anions and positively charged defects, and an inner layer consisting of pure alumina [28]. The concentration profile of these contaminations across the pore walls is inhomogeneous (see, for example [30–32]). Even though it was shown that the pore walls of AAO hard templates are reactive at elevated temperatures of about 500 °C and above [33–35], in the temperature range relevant to the structure formation of soft matter AAO hard templates can be regarded as inert, shape-defining molds.

Disordered, about 60 μm thick AAO membranes with a mean D_p-value of the order of 200 nm are commercially available (Whatman Anopore) [24]. These AAOs feature a broad pore size distribution as well as irregular pore shapes (Fig. 2a). The dispersity of the pore diameter distribution, calculated by dividing the standard deviation by the mean pore diameter, is typically larger than 20%. Initially, Anopore AAOs were designed as filters. Therefore, the nominal pore diameter refers to the narrowest pore segments that determine their separation performance. Moreover, the membrane surface of disordered AAOs exhibits pronounced roughness on different length scales that complicates the removal of residual material after infiltration and the purification of the nanostructures thus obtained.

The two-step MA process reported by Masuda and Fukuda involving self-ordered pore growth can be considered as a major breakthrough in AAO-based nanoprocessing [36]. A first anodization step is carried out in such a way that the initially disordered nanopores self-assemble into a hexagonal lattice. This uppermost alumina layer, the surface of which contains disordered pores, is removed by a selective wet-chemical etching step. The surface of the remaining aluminum substrate is patterned with hexagonal arrays of hemispherical indentations, which are replicas of the pore bottoms of the etched alumina layer. These indentations act as seeds for the growth of a hexagonal array of nanopores with T_p-values of up to several 100 μm in a second anodization step. Three self-ordered MA regimes have been identified. Using sulfuric acid as an electrolyte solution at an anodization voltage of about 25 V yields self-ordered AAO with a lattice constant of 65 nm and a D_p-value of about 25 nm [37]. Anodization with oxalic acid solutions at anodization voltages of about 40 V yields AAO with a lattice constant of 100 nm and a D_p-value of 35 nm [36], and anodization in phosphoric acid solutions at 195 V yields AAO with a lattice constant of 500 nm and a pore diameter of about 180 nm [38, 39]. The pores are arranged in hexagonal lattices charac-

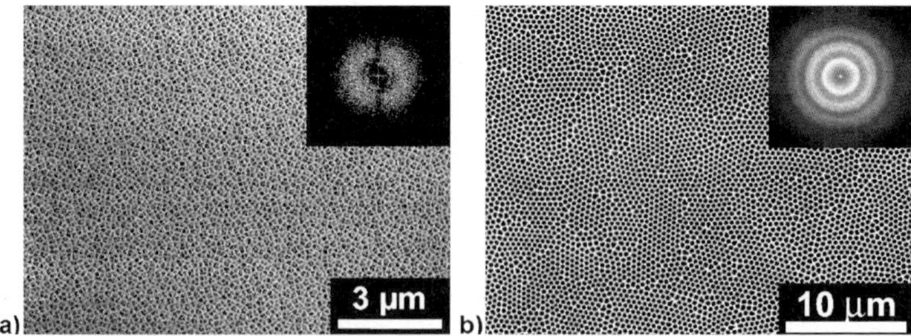

Fig. 2 Anodic aluminum oxide. **a** Example of disordered AAO with a mean D_p-value of 200 nm; **b** self-ordered AAO anodized with a phosphoric acid electrolyte solution under MA conditions. *Insets*: Fourier transforms

terized by a polycrystalline degree of order that consist of grains extending 10 to 20 lattice constants. The pore diameter distributions of self-ordered AAO have a dispersity of less than 8% and are therefore significantly sharper than those of disordered AAO. It was noted that the pore arrays produced in the self-ordering MA regimes exhibit a porosity of 10% [40]. Porosities up to 50% can be achieved if the pores are widened by isotropic wet-chemical etching (Fig. 2b). Self-ordered AAO produced by mild anodization is attached to underlying aluminum substrates and therefore mechanically stabilized. Moreover, the surface of self-ordered AAO is significantly smoother than that of disordered AAO. These two features substantially facilitate the fabrication, purification and characterization of nanoobjects prepared inside the pores of such hard templates, as well as the fabrication of functional membranes. Selective etching steps can be applied to remove the aluminum substrate and to open the pore bottoms.

MA requires several days of processing time and self-ordering pore growth occurs only in narrow process windows. Recently, the so-called hard anodization (HA), which is routinely being used in industrial processes, has been investigated as a new and complementary access to AAO hard templates. HA is performed at higher anodization voltages than MA, and the alumina layers grow orders of magnitude faster. Lee et al. reported that HA with oxalic acid solutions at anodization voltages between 120 and 150 V yields AAO with D_p-values of 49–59 nm and lattice constants of 220–300 nm,

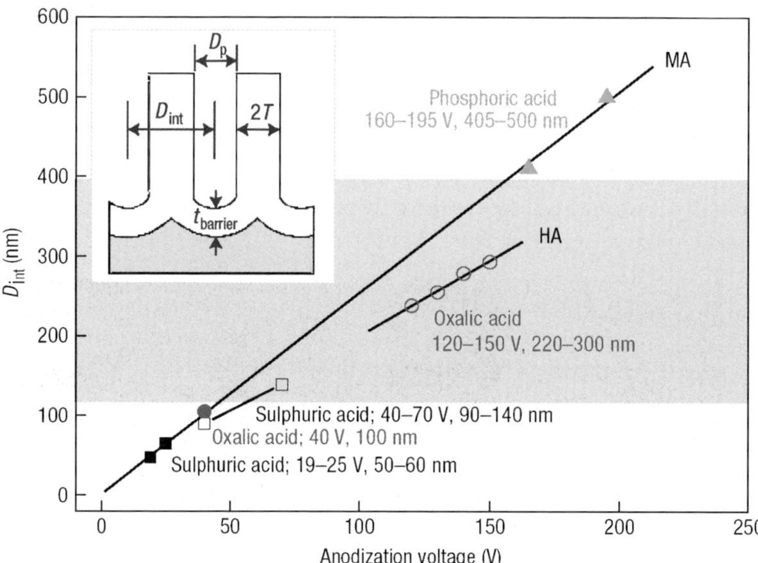

Fig. 3 Overview of currently identified self-ordering MA and HA regimes for the production of AAO. Reproduced from [41]. © (2006) Nature Publishing Group

a range not covered by the MA self-ordering regimes [41]. Moreover, the initial porosity of the HA membranes obtained with oxalic acid solutions lies in the range from 3.3 to 3.4% and is therefore significantly smaller than that of MA membranes. This is an important property of HA membranes that could enable the production of mechanically stable nanofiber arrays in which large distances between the nanofibers are required to prevent them from condensation. Recent efforts to conduct anodization of aluminum in sulfuric acid solutions under HA conditions to produce AAOs with lattice constants below 100 nm suffer from the poor mechanical stability of the AAOs thus obtained [42]. An overview of currently identified self-ordering MA and HA regimes is given in Fig. 3. Another important progress reported by Masuda and co-workers is the high-temperature anodization of aluminum in concentrated sulfuric acid solutions yielding self-ordered AAO with lattice constants as small as 30 nm and D_p-values of 18 nm [43, 44]. Long-range ordered AAO is accessible by combining self-ordering MA or HA regimes with lithographic prepatterning of the aluminum substrates used for anodization by means of hard-imprint lithography. However, this approach is limited to lattice constants larger than 100 nm because no master stamps with a smaller feature size are available [41, 45–47].

3
Nanotubes by Infiltrating Nanoporous Hard Templates

3.1
Wetting: Basic Concepts

The starting point for the fabrication of nanotubes by means of nanoporous hard templates is the infiltration of the target materials, of precursors thereof, or of monomers, into the pores. This process is far from being trivial and only partially understood. However, it is reasonable to assume that interfacial interactions dominate, or at least significantly influence both the infiltration and the mesoscopic structure formation. In the following, we will assume that exclusively physisorption occurs at the interface between the pore walls and the infiltrated material, neglecting the possibility of specific chemical interactions. In the literature it is often assumed that the filling of nanopores is driven by capillary action. As discussed below, this is only the case under certain conditions. To gain a qualitative understanding of the physico-chemical phenomena underlying the penetration of liquids into nanochannels, at first the well-investigated microscopic behavior of fluids deposited on smooth substrates will be discussed briefly. The surface energy of the substrate, the surface tension of the liquid, and the liquid/solid interfacial energy determine the macroscopic contact angle, at which a liquid/vapor interface meets the solid surface, according to Young's law [48]. The equilibrium contact angle is

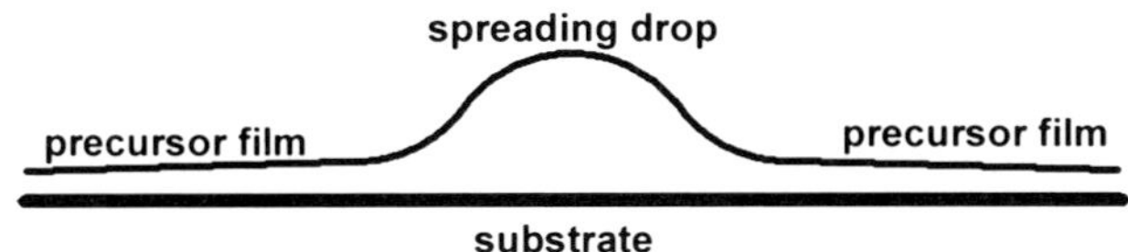

Fig. 4 Schematic diagram of a liquid drop spreading on a smooth substrate

a common measure of the wettability of a solid surface by a specific liquid. The system will adopt a state where the overall interfacial energy is minimized. A zero contact angle is equivalent to the spreading of the liquid, that is, to the maximization of the liquid/solid contact area. Commonly, polar inorganic surfaces exhibit high surface energies, whereas those of organic liquids and polymeric melts are about one order of magnitude lower [49]. Consequently, organic fluids commonly spread on inorganic, oxidic substrates. If the surface energies of the solid and the liquid converge, the contact angle will be larger than zero.

The question arises as to how a drop of a low-energy liquid spreads on a high-energy surface. Even for viscous polymeric fluids the formation of a precursor film could be evidenced [50–52]. At the foot of the drop, where the liquid contacts the solid surface, a thin film of the liquid emanates and covers large areas of the substrate (Fig. 4). In the proximity of the drop, the thickness of the film is in the mesoscopic range, whereas at the spreading front the film is thinner than a monolayer, as determined by ellipsometry, indicating an incomplete surface coverage. As the precursor film proceeds, the height of the macroscopic drop decreases. Taking into account that a finite amount of the fluid spreads on a surface large enough to be considered as "infinite", a "pancake" structure is nevertheless to be expected, that is, a liquid film covering a finite area with a thickness exceeding that of a monolayer. This is because the interactions between the solid and the liquid comprise attractive long-range interactions [53–55] expressed in terms of the so-called "disjoining pressure", that is, the pressure that has to be exerted to prevent the liquid film from thickening.

3.2
Precursor Wetting of Porous Templates with Polymeric Melts

When a fluid spreads on the walls of a nanochannel with a finite length, the situation is different from the spreading on smooth substrates in that the presence of an infinite bulk reservoir of the liquid can be assumed, whereas the solid surface to be wetted is finite. The infiltration of a liquid into an empty pore is qualitatively similar to the replacement of a liquid phase filling a cylindrical channel by another one having higher affinity to the walls of the channel. This process was intensively investigated theoretically and experimentally [56–59] because of its practical relevance to oil recovery from bituminous sands, into which aqueous solutions are injected to replace and extract the oil. The underlying physics is complex and dependent on the dimensions of the channels hydrodynamic phenomena or interfacial phenomena dominate. In brief, at first a wetting film consisting of the liquid infiltrating the pores covers the pore walls (Fig. 5a). Instabilities in this film may occur (Fig. 5b), and as more and more liquid moves into the pores, these instabilities begin to grow until a "snap off" or "pinch off" takes place, that is, a meniscus forms (Fig. 5c). The interfaces of the meniscus move in opposite directions, and the pore volume is eventually completely filled with the liquid (Fig. 5d). It is reasonable to assume that the "snap-off" mechanism generally guides the infiltration of nanoporous and microporous materials with fluids.

If liquid, disordered polymers are brought into contact with porous hard templates exhibiting high surface energy and D_p-values significantly exceeding twice the radius of gyration of the infiltrated polymer, a polymeric precursor film with a thickness of a few tens of nanometers rapidly covers the entire area of the pore walls on a time scale of seconds to minutes even if the pores have a depth of the order of 100 micrometer. This behavior is commonly observed for AAO hard templates. Despite the fact that the pores should be completely filled in equilibrium, the liquid polymer layer is stable at least for several days. Consequently, polymer nanotubes can be obtained by solidifying the polymer [60–63]. For example, Fig. 6 shows a broken nanotube consisting of PS ($M_n \approx 850\,000$ g/mol) protruding from a self-ordered AAO template with a D_p-value of 400 nm and a T_p-value of 100 μm.

Up to now, the formation of mesoscopic polymer layers on the pore walls and the reasons for their kinetic stability are only qualitatively understood, if at all. It appears that the macromolecules have to be mobile enough to be removed from the bulk and to diffuse into the pores. Both entropic relaxation of the polymer chains and the disjoining pressure, as discussed in the

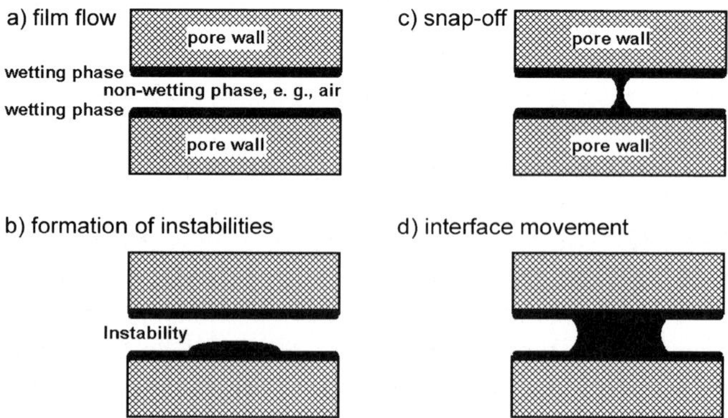

Fig. 5 Infiltration of a low molecular weight liquid (*black*) into a cylindrical channel

Fig. 6 PS tube prepared by precursor wetting protruding from an AAO hard template. Reproduced from [60]. © (2002) AAAS

previous section, may contribute to the generation of a polymer layer having a mesoscopic thickness. When the diameter of the pores in the hard template is reduced below about twice the radius of gyration of the infiltrated polymeric species, the hollow space in the tubes disappears and solid rods are obtained [64–67]. In general, precursor wetting takes place if the pore walls of the hard template exhibit high surface energy and if the polymeric melts are heated to temperatures well above their glass transition temperature [61, 68, 69]. However, many aspects of precursor wetting still need to be elucidated. Little is known about the relaxation processes leading to the formation of the mesoscopic polymer film, about the conformation of the polymeric chains in the nanotube walls, and about the parameters influencing the thickness of the nanotube walls.

3.3
Capillary Wetting of Porous Templates

If polymeric melts are infiltrated into nanoporous hard templates under conditions where the formation of a precursor film is suppressed, the filling of the pore volume is governed by classical capillarity, a mechanism that was intensively investigated in the context of capillary molding [70, 71]. The strong adhesive forces between the polymer and the pore walls are still effective but are not strong enough to drive single molecules out of the polymeric bulk reservoir on top of the hard template. However, removing single molecules from the bulk is a prerequisite for the rapid formation of a precursor film. In equilibrium, the overall interfacial energy is nevertheless minimized by completely filling the pore volume with the polymeric melt. To reach the equilibrium, a solid cylinder of the liquid but viscous polymer, preceded by a meniscus, slowly moves into the pores of the hard template [72] until the entire pore volume is filled. For example, Fig. 7a displays an array of nanofibers consisting of PS-b-PMMA infiltrated at 200 °C into an AAO membrane ($D_p = 400$ nm, $T_p = 100$ µm). The menisci of the nanofibers are clearly

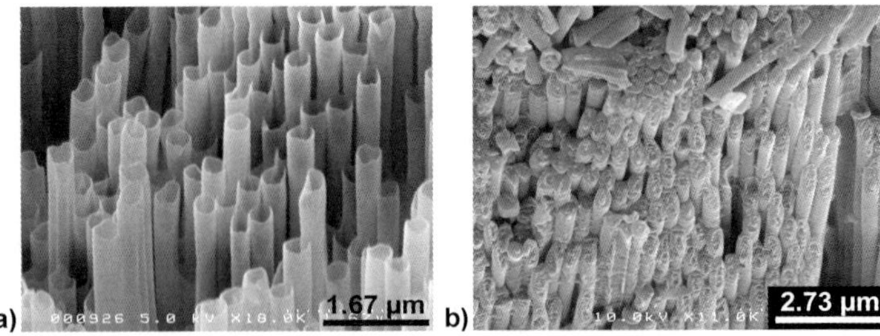

Fig. 7 PS-*b*-PMMA nanofibers prepared by capillary wetting into AAO hard templates. **a** Tips with menisci; **b** section through a PS-*b*-PMMA nanofiber array, evidencing the solid rod-like nature of the nanofibers. The SEM images are a courtesy Dr. Olaf Kriha

seen. The investigation of specimens containing fiber arrays cut along a plane slightly inclined with respect of the fiber axis reveals the solid nature of the fibers (Fig. 7b) [73].

The length of the fibers is proportional to the square root of the infiltration time, that is, the time the polymer is kept in the liquid state while in contact with the hard template [61, 72–74]. The time scale on which the pores are filled conveniently allows adjusting the length of the polymeric fibers by quenching the infiltrated polymer below its glass transition temperature or its crystallization temperature, respectively. For example, Moon and McCarthy could adjust the lengths of PS fibers prepared by melting PS ($M_W = 280\,000$ g/mol) at 200 °C in contact with an AAO membrane having a pore diameter of 200 nm to 0.6, 0.9, 1.2 and 1.6 µm by heating the polymer for 5, 10, 15 and 20 minutes, respectively [72]. Kriha et al. reported that loading a BCP melt with weight accelerated the infiltration, and that template pores ($D_p = 400$ nm) with a T_p-value of 100 µm were completely filled after 6 h [73].

Microphase-separated BCPs commonly fill the pores of hard templates via capillary wetting (Fig. 7) [73–75]. This is to be expected since the removal of single molecules from the bulk would disturb the ordered structure in the BCP. Moreover, the blocks had to diffuse (or to drift) through domains consisting of the other component. Thus, they had to overcome repulsive enthalpic interactions. In the case of disordered homopolymer melts, apparently a transition from precursor wetting to capillary wetting occurs that appears to be related to an increase in the viscosity of the polymeric melt, if hard templates having pore walls with a high surface energy are used. For example, PS ($M_n = 30\,500$ g/mol) forms solid rods in AAO membranes with a D_p-value of 200 nm after annealing for 2 h at 130 °C. However, increasing the infiltration temperature to 205 °C resulted in the instantaneous formation of nanotubes with lengths of 60 µm, corresponding to the T_p-value of the template pores [61]. She et al. observed that wetting AAO membranes having a D_p-value of 200 nm with polyamide 66 at 250 °C yielded nanowires [68].

Similar results were obtained for poly(propylene) [69]. Correspondingly, at a given infiltration temperature and for a given polymer, precursor wetting will occur if the molecular weight of the polymer is relatively low, whereas capillary wetting will occur in the case of relatively high molecular weights. For example, PS with a M_n-value of 30 500 g/mol forms nanotubes when infiltrated into AAO with a D_p-value of 200 nm at 205 °C, as discussed above. However, if the PS has a molecular weight of about 760 000 g/mol, again short nanorods where obtained [61]. It was suggested to exploit the dependence of the infiltration mechanism on the molecular weight for fractionating polymers with different molecular weights [61]. It should be noted that precursor wetting and capillary wetting represent only different kinetic routes to the equilibrium that is characterized by complete filling of the pores with the polymer. In the case of hard templates having pores with D_p-values of a few tens of nm, not only capillary wetting but also precursor wetting will yield solid nanorods, as discussed in the previous section. However, the rates at which the pores are filled with the polymer potentially allow for distinguishing between both mechanisms. Shin et al. reported a significantly enhanced mobility of polymer chains in very narrow pores having D_p-values smaller than the radius of gyration of the infiltrated polymer, a finding that was attributed to a confinement-induced decrease in the degree of intermolecular entanglement [76]. Because of the confinement imposed by the pore geometry, the chains are not stretched in the direction of flow but are compressed in a direction orthogonal to the flow. Consequently, it is reasonable to assume that the parameter that determines the wetting mechanism is the energetic and entropic effort required to remove the polymer chains from the bulk reservoir on top of the hard template and to draw them into the pores.

Combinations of both wetting mechanisms identified so far allow fabricating new and unprecedented one-dimensional nanostructures, for example, tube/rod hybrid fibers. To this end, AAO hard templates were infiltrated with BCPs under conditions of capillary wetting in such a way that pore segments with an adjustable length were filled with solid BCP rods. Then, the hard templates were turned upside down and a homopolymer was infiltrated from the reverse side under conditions of precursor wetting. The composite fibers thus obtained consisted of a stiffer, solid BCP segment and a more flexible, tubular homopolymer segment (Fig. 8) [73].

Another means to control the infiltration mechanism is the surface energy of the pore walls of the hard template. For example, Grimm et al. obtained short polymer rods with hard templates modified with a silane coupling agent bearing perfluorated organic moieties under conditions where otherwise tubes had formed [20]. Thus, the formation of a precursor film can be suppressed completely by reducing the surface energy of the hard template. However, under these conditions, the polymeric melt might not spread on

Fig. 8 Cross-sectional view of an array of tube/rod hybrid nanofibers obtained by infiltrating polymers into hard templates at first under conditions of capillary wetting and subsequently from the reverse side of the hard template under conditions of precursor wetting. Reproduced from [73]. © (2007) Wiley-VCH

the pore walls and external pressure has to be applied to inject the molten polymer into the pores.

3.4
Template Wetting with Polymeric Solutions

The infiltration of solutions consisting of a polymer and a volatile low molecular mass solvent into nanoporous hard templates has been intensively investigated [13, 65, 68, 69, 77–81]. In principle, it is sufficient to drop the solution onto the hard template and to let the solvent evaporate. However, it is far from being trivial to predict whether the properties of such a mixture rather correspond to those of a low-molecular mass liquid or to those of a polymeric melt. Two limiting cases can be postulated: If the solution at first completely fills the pore, the polymeric layer on the pore walls forms by adsorption from solution. In this case, the concentration of the polymer and the volume of the solution deposited on the hard template will certainly affect the formation of the polymeric nanostructures inside the pores. The solvent evaporates at the solvent/air interface so that the solution becomes more and more concentrated. The increasing concentration of the polymer will change the nature of the adsorbed polymer layer on the pore walls. Eventually, it will become important which portion of the polymer contained in the applied volume of the stock solution is still located on top of the hard template when the polymer solidifies as the solvent content drops. If the mixture behaves like a polymeric melt, it is to be expected that it will infiltrate the pores according to the precursor wetting mechanism since the solvent acts as a plasticizer. Then, a swollen mesoscopic polymer layer covers the pore walls. In any case, the evaporation of the solvent, a process that can hardly be controlled in a satisfying manner, will influence the morphology of the polymeric nanostructures. Liquid/liquid phase separation [82, 83] and wetting transitions [84] may occur when the composition of the system changes. Moreover, evaporation may lead to a temporary concentration gradient inside the pores along with non-uniform vitrification [85]. Structure and density of the absorbed layer will not only strongly depend on the polymer/solvent interactions [86–88] but also on environmental conditions such as temperature and humidity. Little is known about the conformation of polymer chains covering the pore walls of hard templates infiltrated by polymeric solutions. Primak et al. studied PDMS films ($M_w = 10\,940$ g/mol) deposited from a solution in chloroform into AAO membranes by deuterium nuclear magnetic resonance spectroscopy [80]. They found a high degree of surface-induced ordering inconsistent with the expected loop/tail conformations and suggested that the chains in the proximity of the pore walls were flattened and that particularly strong interactions between the monolayer covering the pore walls and the pore walls were present. However, it remains unclear to what extent these findings are specific to the system investigated by these authors.

In contrast to the conformation of the polymer chains, the morphology of polymeric nanostructures can easily be probed by SEM and TEM. Ai et al. reported that nanotubes are obtained if diluted solutions of PS ($M_w = 270\,000$ g/mol) in cyclohexane infiltrate AAO membranes at 35 °C, that is, under θ-conditions, whereas infiltration of concentrated solutions results in the formation of solid rods [81]. This is in line with results reported by Song et al., according to which the wall thickness of PS nanotubes deposited from solutions in dichloromethane increased with increasing concentration of PS [77]. Chen et al. obtained amorphous carbon nanotubes by infiltration of solutions of PAN in DMF, crosslinking of the PAN and subsequent carbonization. Again, the walls of the so-obtained carbonaceous nanotubes were thicker for higher PAN concentrations in the infiltrated stock solutions. In this case, the wall thickness could also be tuned by performing successive infiltration-pyrolysis cycles [89]. However, it should be noted that infiltration of solutions into AAOs often results in the formation of short, defect–rich fiber segments [79].

Only few attempts have been made to rationally design the mesoscopic fine structure of nanotubes fabricated by wetting nanoporous hard templates with polymeric solutions. Chen et al. infiltrated solutions of PS-*b*-PAN in DMF into AAO. As described above, the PAN was at first crosslinked and then carbonized. However, the PS domains were converted into holes, and porous amorphous carbon nanotubes could be fabricated [89]. In a similar approach, Rodriguez et al. used a solution of PS-*b*-PVP and carbohydrates associated with the PVP blocks via hydrogen bonds in DMF into AAO and obtained mesoporous amorphous carbon nanotubes with the positions of the mesopores determined by the positions of the PS domains. Solvent annealing of the BCP/hydrocarbon films in DMF/benzene vapor led to a significantly

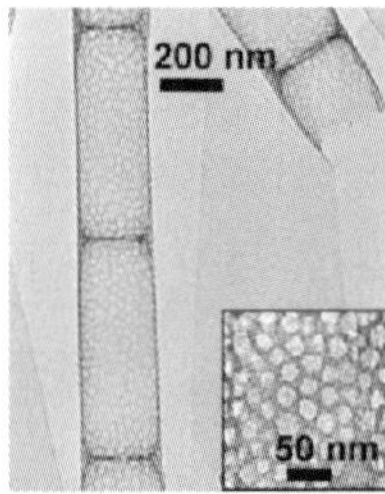

Fig. 9 TEM image of nanoporous carbonaceous nanotubes prepared using PS-*b*-P2VP with a bamboo-like structure. *Inset*: Hexagonal arrays of pores on the tube wall. Reproduced from [90]. © (2006) American Chemical Society

more uniform distribution of the PS domains and hence of the pores in the amorphous carbon nanotubes [90]. Apparently, in contrast to the infiltration of polymeric melts, tubular structures with walls consisting of BCPs can be obtained in this way. In both works, the hard AAO templates had a D_p-value of about 200 nm, and the occurrence of bamboo-like morphologies or ring-like ribs arranged more or less periodically along the nanotubes was reported (Fig. 9). Chen et al. attributed the formation of this structure to self-organization processes related to the evaporation of the solvent, similar to those reported by Gonuguntla and Sharma, who investigated the evaporation of an initially pure solvent drop on a smooth, dissolving polymer substrate [91].

An interesting self-ordering phenomenon is the occurrence of Rayleigh-Plateau instabilities. It is well known that annular liquid films are, similar to liquid cylindrical threads, susceptible to the growth of periodic thickness fluctuations [92, 93]. Chen et al. reported that nanotubes prepared by infiltrating AAO hard templates with 5 wt-% solutions of PMMA ($M_w = 22\,700$ g/mol) in chloroform can be converted into nanorods with periodic encapsulated holes driven by the Rayleigh instability [94]. At first, a smooth polymer film covered the pore walls. Thermal annealing of the PMMA/AAO hybrid membrane at temperatures above the glass transition temperature of the PMMA resulted in the growth of thickness undulations in the annular PMMA film and finally in the formation of bridges across the entire nanopore. The wavelength of the periodic structure increases with D_p and amounts to about 1000 nm for a D_p-value of 200 nm. Figure 10a shows a TEM image of a PMMA tube with periodically undulated pore walls, and Fig. 10b a TEM image of a hole-containing PMMA nanorod.

On the one hand, template wetting with polymeric solutions is, up to now, poorly understood and difficult to perform in a reproducible manner under well-defined conditions. On the other hand, solution wetting offers a plethora of possibilities to generate one-dimensional nanostructures exhibiting complex morphologies and fine structures. To this end, as discussed in Sect. 5, mixtures of functional materials or precursors thereof in a common solvent

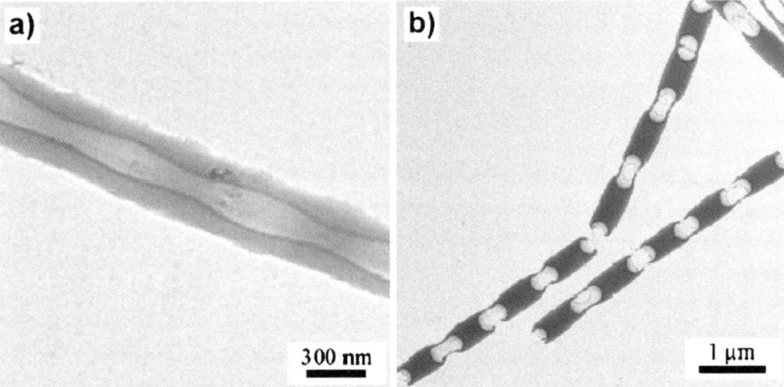

Fig. 10 Rayleigh instabilities in PMMA nanofibers. **a** TEM image of a PMMA tube with periodically undulated pore walls; **b** TEM image of a hole-containing PMMA nanorod. Reproduced from [94]. © (2007) American Chemical Society

can be infiltrated into hard templates to exploit phase separation processes inside the pores as a structure formation process.

4
Self-Assembly of Single-Component Materials in Nanopores

4.1
Overview

Crystallization and mesophase formation determine the optical, electronic, mechanical, chemical and piezoelectric properties of polymeric materials to a large extent. For example, supramolecular self-organization enhances charge transport in conjugated polymer semiconductors [95]. Also, the performance of stimuli-responsive materials may depend on their mesoscopic fine structure. Whereas in inorganic ferroelectrics slight dislocations within the crystalline unit cell lead to polarization switching, changing the orientation of the polar axis in ferroelectric polymers [96] is associated with rotations about C – C bonds in the backbone [97–99]. Consequently, the polar axis of the unit cells in ferroelectric polymer crystals can only adopt certain orientations which are coupled with the crystal orientation. It is, therefore, reasonable to assume that control over crystallization and mesophase formation inside hard templates is crucial to the rational design of one-dimensional nanostructures.

Little is known about the nature of disordered, amorphous interphases covering the pore walls of nanoporous hard templates, but a limited body of literature deals with ordered supramolecular architectures in single-component nanotubes and nanorods confined to hard templates. Mesoscopic morphologies characterized by a crystalline or liquid-crystalline degree of order and by pronounced anisotropy can be generated by infiltration of semicrystalline or liquid-crystalline polymers as isotropic liquids and subsequent crystallization or mesophase formation inside the hard template pores. Common experimental techniques, including WAXS [64, 100], SAED [65, 101], HRTEM [102, 103], polarized IR spectroscopy [7, 10, 12, 101], and DSC [64, 100, 104] have been applied to characterize ordered supramolecular architectures in one-dimensional nanostructures. WAXS on ensembles of aligned nanofibers provides valuable information on textures, whereas SAED and HRTEM can be used to locally elucidate crystal orientations and to determine the polymorph formed. Whereas WAXS is easy and straightforward to apply on AAO membranes loaded with soft matter, electron microscopy in general suffers from strong interactions between the nanostructures being probed and the incident electron beam. This drawback may be overcome by short exposure times or by cooling the samples [65]. Up to now, SAXS is no established method for probing textures and mesoscopic features such as long periods in one-dimensional nanostructures. As long as they are aligned within the templates, the background scattering of the matrix material may pose problems for the evaluation of the SAXS patterns, and in the case of released nanofibers the inhomogeneous nature of the powders used for performing SAXS experiments hampers the analysis of the collected data. Much information can be gained by polarized IR spectroscopy [105]. On the one hand, it is often possible to assign specific peaks to amorphous and crystalline material so that the crystallinity can be deduced from the peak areas. On the other hand, anisotropy is obvious from infrared dichroism that is accessible by comparing peak areas in spectra taken with IR beams whose planes of polarization are inclined by specific angles. DSC yields information on the crystallization kinetics and nucleation mechanisms, as well as on crystal sizes. In principle, NMR and dielectric spectroscopy should also be applicable methods that have, up to now, virtually not been explored for the study of supramolecular architectures in nanoporous hard templates. However, at least for NMR, a proof of concept was reported by Primak et al. [80].

Mesoscopic structure formation processes inside hard templates can be influenced by surface-induced ordering and geometric confinement since their characteristic length scales are of the same order of magnitude as the D_p-values of the nanopores or the thickness of the nanotube walls they contain. Already early works on template syntheses of functional polymers by Martin and co-workers indicated that supramolecular order and properties such as conductivity may be enhanced inside hard templates [11]. Polypyrrole and poly(3-methylthiophene) [6], polyacetylene [7] and polyaniline [10] nanofibers having diameters in the mesoscopic range were obtained by synthesizing the corresponding polymers in the nanopores. Their enhanced conductivity, which was evidenced for nanofibers aligned in the hard templates and for mats consisting of released nanofibers [106] was attributed to enhanced supramolecular order [7, 11, 12, 107]. Polymerization within the pores involves preferential growth of the chains located on the pore walls of the hard template because of their reduced solubility as compared to the monomer. Therefore, after short polymerization times, the nanotubes thus formed predominantly consist of polymer chains in proximity to the pore walls, where they are aligned and the formation of kinks and bends is suppressed. Martin and co-workers therefore concluded that the enhanced supramolecular ordering thus realized is accompanied by an increased conjugation length, which in turn results in higher conductivity. This surface-induced alignment was found to decay when thicker nanotube walls were prepared by longer polymerization times.

4.2
Crystallization of Thermoplastics

One of the most important structure formation processes in polymeric materials having chain architectures that allow, at least to some extent, packing of the chains is crystallization. The crystallinity of semicrystalline polymers, as well as the morphology and the orientation of the crystalline entities, largely determine the properties of these materials. Polymers usually crystallize as lamellar crystals in which folded chains are oriented approximately perpendicular to the surface of the lamellae [108–110]. The typical thickness of these crystals lies in the nanometer range, while their lateral dimensions are in the micrometer range, thus by far exceeding typical D_p-values of hard templates. Within the crystals, the chains adopt a helical conformation, and the growth of the lamellae proceeds in the lateral directions. On a larger scale, the lamellae are organized in spherulites, densely branched, isotropic, polycrystalline superstructures [111–113]. It is an intriguing question as to how the geometric confinement of the pores in nanoporous hard templates and their large surface-to-volume ratio influences the crystallization of polymers. In the case of melt infiltration of semicrystalline polymers into hard templates, crystallization is an important issue because crystallization may occur upon cooling to room temperature. Even though, up to now, only a limited number of publications deals with this topic [64, 79, 100, 101, 104], it has become clear that crystallization of polymers confined to the pores of hard templates can be influenced, and to some extent engineered, by the presence or absence of a bulk polymer reservoir in contact with the polymer inside the pores, by the D_p-value of the hard template, and by the temperature profile applied.

Generally, the c-axis of the polymeric crystals, which is commonly normal to the plane of the lamella crystals, orients perpendicular to the pore axes. This enables the lamellae to grow along the pores. Moreover, the crystallinity of the material inside the pores is typically below that of the corresponding bulk material. In the case of non-isothermally crystallized PVDF, the crystallographic direction exhibiting the highest growth rate, that is, the $\langle 020 \rangle$ direction, aligns with the pore axis, resulting in uniform crystal orientation inside the pores on a macroscopic scale [79], if crystallization is initiated by heterogeneous nucleation [114, 115] in a bulk PVDF reservoir on top of the hard template. The lamellae in the spherulites in the bulk reservoir are oriented in such a way that their direction of fastest growth points radially outwards. If a growing spherulite hits the surface of a hard template infiltrated with PVDF, only those lamellae proceed into the pores whose $\langle 020 \rangle$ direction is, within a certain tolerance, oriented parallel to the pore axes [64]. If the bulk reservoir is removed from the surface of the hard template, crystallization in each crystallizing entity is predominantly initiated by homogeneous nucleation at high supercooling, because the probability of the occurrence of heterogeneous nuclei in the small volume of the nanopores is negligible. Then, crystals with one of the $\langle hk0 \rangle$ directions aligned with the pore axes, a crystal orientation that enables growth of the lamellae along the pores, form with statistical frequency. This was evidenced by WAXS measurements performed in $\Theta/2\Theta$ geometry on PVDF nanofibers aligned within the pores of the hard templates, crystallized in the presence or absence of a bulk PVDF reservoir. In the former case, only the (020) peak appears in the XRD pattern (Fig. 11a), in the latter case all $(hk0)$ peaks show up (Fig. 11b) with relative intensities similar to those in the powder pattern of isotropic PVDF (Fig. 11c). However, it is striking that reflections with non-zero l-index are still missing in the pattern of the sample crystallized in the absence of the bulk reservoir. This is because crystals with a corresponding orientation impinge on the pore walls and therefore occupy only a small portion of the pore volume. DSC cooling scans nicely confirmed that homogeneous nucleation significantly contributes to the crystallization

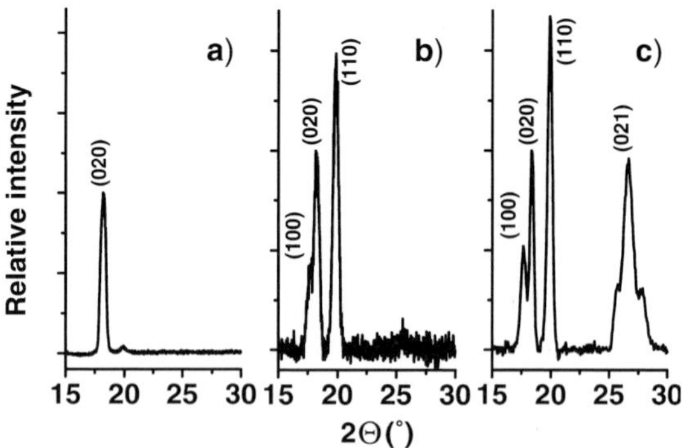

Fig. 11 WAXS patterns of PVDF nanofibers aligned in an AAO hard template ($D_p = 35$ nm) measured in $\Theta/2\Theta$ geometry. **a** Non-isothermal crystallization in the presence of a bulk PVDF surface reservoir; **b** non-isothermal crystallization in the absence of a bulk PVDF surface reservoir; **c** powder pattern of an isotropic sample

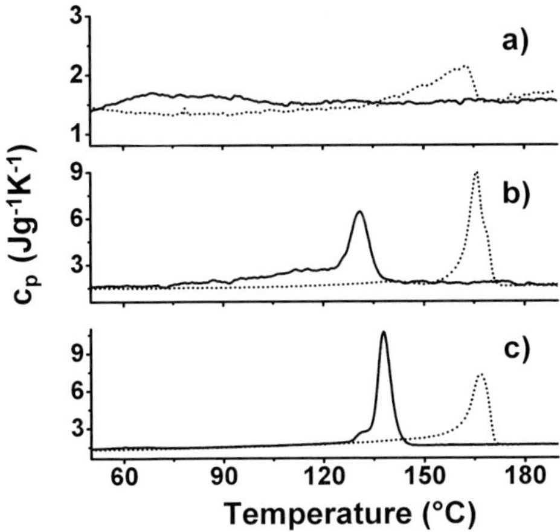

Fig. 12 DSC scans of separated PVDF nanostructures within templates (*dotted curves*: heating runs; *solid curves*: cooling runs; heating and cooling rates: 20 K = min). **a** Nanorods (D_p = 35 nm); **b** nanotubes (D_p = 400 nm); **c** bulk PVDF. The *curves* were corrected by subtracting the contribution of the alumina (determined by reference measurements of empty templates). Reproduced from [64]. © (2006) American Physical Society

of an ensemble of separated PVDF nanotubes inside an AAO template with a D_p-value of 400 nm, whereas exclusively homogeneous nucleation occurs if the D_p-value is reduced to 35 nm (Fig. 12). Random PVDF copolymers with trifluoroethylene P(VDF-*ran*-TrFE) infiltrated into AAOs with D_p-values ranging from 55 to 360 nm were also investigated and found to be crystalline. By probing the relative permittivity of arrays of P(VDF-*ran*-TrFE) nanofibers the ferroelectric-to-paraelectric phase transition could be observed [66, 67].

Woo et al. investigated the crystallization kinetics of separated entities of linear PE inside AAO hard templates by DSC and came to the conclusion that inside pores with D_p-values below about 50 nm heterogeneous nucleation at the pore walls becomes dominant, whereas for D_p-values of 62 and 110 nm homogeneous nucleation initiates crystallization [104]. The Avrami constant n that depends on the growth geometry and the nucleation mechanism was found to be smaller in the case of PE crystallizing in the nanopores than for bulk crystallization, indicating that crystal growth inside the pores is frustrated and dominated by nucleation occurring at high supercooling. Plotting the reciprocal crystallization half time, that is, the crystallization time at which the crystallinity reaches half of the finally attained value at a given crystallization temperature, versus the crystallization temperature T_c and the supercooling, respectively, revealed that the crystallization rates for bulk samples depend on the degree of supercooling. This dependence was significantly more pronounced for PE confined to AAO hard templates hav-

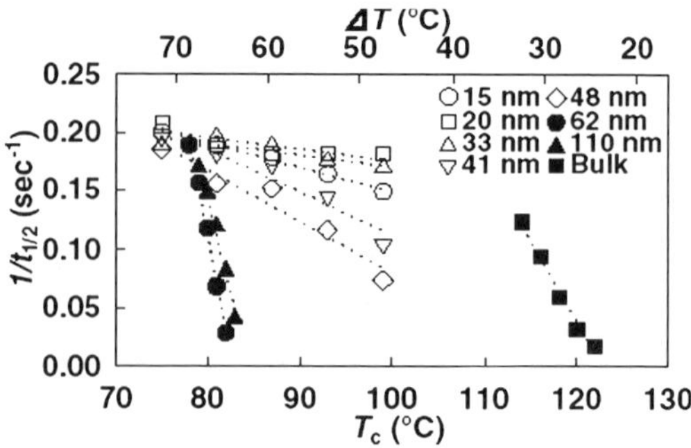

Fig. 13 Reciprocal of crystallization halftime ($1/t_{1/2}$) of PE confined to AAO with different D_p-values as a function of the crystallization temperature (T_c) and the degree-of-supercooling (ΔT). Reproduced from [104]. © (2007) American Physical Society

ing D_p-values of 110 and 62 nm. However, for D_p-values below 50 nm, only a weak dependence of the crystallization rate on the supercooling was found (Fig. 13), indicating the transition from dominant homogeneous nucleation in the larger pores to heterogeneous surface nucleation in the smaller pores. This transition was attributed to the fact that the smaller pores have a much larger surface-to-volume ratio, whereas the frustration of the crystal growth is more pronounced too.

A more detailed analysis of WAXS patterns of PE inside AAO hard templates revealed that, even though the ⟨110⟩ direction is being considered as the direction of fastest crystal growth, the b-axis aligns with the long axes of the template pores [100]. Whereas the orientation distribution of the crystals was found to be narrower in smaller pores, DSC melting runs also revealed a shift of the position of the melting endotherms to lower melting temperatures T_M along with smaller D_p-values. Consequently, the calculation of T_M using the Gibbs–Thomson equation, assuming that crystal growth is confined in three dimensions, yielded results fitting much better to the experimental values than T_M-values calculated assuming unlimited crystal growth along the pores. Also, the crystallinity decreased with decreasing D_p. Shin et al. concluded that crystal growth of PE along the pores is indeed restricted because the dominant growth direction of the crystals is not aligned with the long axes of the template pores.

Polarized IR spectroscopy was employed by Wu et al. to investigate crystallinity, the formation of different crystal orientations and the crystal texture of syndiotactic PS (sPS) crystallized inside AAO hard templates while in contact with a bulk reservoir of the same polymer [101]. The β-polymorph was obtained by cooling from the melt to 260 °C and crystallizing at this temperature for 2 h, while heating amorphous samples quenched from the molten state to 240 °C and heating to this temperature for 2 h resulted in the formation of the α-polymorph. A comparison of the areas of peaks

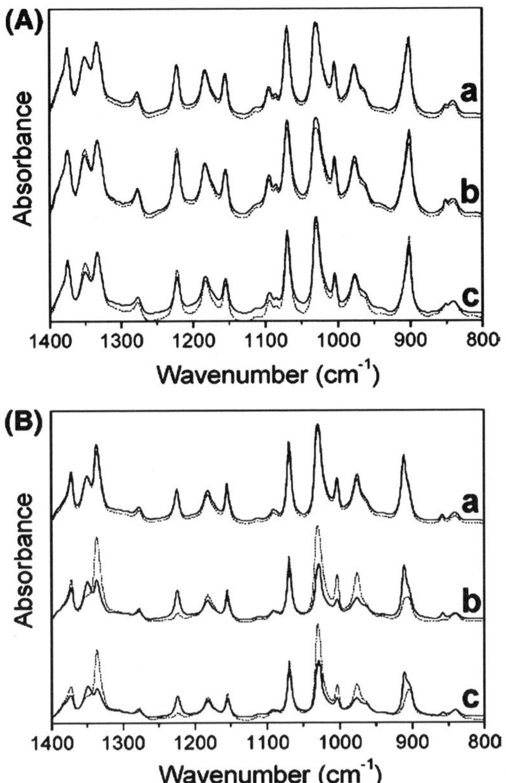

Fig. 14 Polarized infrared spectra of sPS crystallized at lower temperatures (**A**) and at 260 °C (**B**). (*a*) represents the bulk, (*b*) nanorods prepared inside AAO with $D_p = 200$ nm, and (*c*) nanorods prepared inside AAO with $D_p = 80$ nm. Polarization perpendicular to nanorod axes: *solid lines*; polarization parallel to nanorod axes: *dashed lines*. Reproduced from [101]. © (2007) American Chemical Society

characteristic of amorphous sPS and the β-polymorph revealed again a decreasing crystallinity along with decreasing D_p. For the bulk and D_p-values of 200 and 80 nm, crystallinities of 62.0, 49.8 and 36.2%, respectively, were obtained. The evaluation of polarized IR spectra measured on bulk sPS and sPS confined to AAOs with D_p-values of 200 and 80 nm (Fig. 14) revealed that samples crystallized at 240 °C consisting of the α-polymorph were isotropic. In samples crystallized at 260 °C consisting of the β-polymorph no preferred orientation of bulk sPS was observed, whereas inside the AAO hard templates the c-axes in the crystals and thus the chain axes were oriented perpendicular to the pore axes. The apparent differences in the degree of crystal orientation in the AAO/sPS hybrid samples were attributed to the different thermal histories. The α-polymorph was obtained by heating quenched amorphous samples from low temperatures to the target crystallization temperature of 240 °C. Therefore, a temperature range was passed that is characterized by a high nucleation rate at high supercooling. Therefore, crystallization was governed by nucleation, and because of the presence of many small growing crystallites no texture could develop. However, the β-polymorph was obtained by cooling from the isotropic melt in contact with a bulk reservoir. Then, it is to be expected that crystallization is initiated by a small number of heterogeneous nuclei so that crystallization is dominated rather by crystal growth than by nucleation. Consequently, crystals having a major growth direction oriented parallel to the pore axes dominate.

O'Carroll et al. infiltrated molten poly(9,9-dioctylfluorene) (PFO) at 250 °C into AAO membranes and obtained solid nanorods consisting of this polymer [116]. As compared to the bulk PL spectrum, a red-shift of the peaks along with an enhanced vibronic structure was found in the PL spectrum of the nanorods, indicating a narrowed orientation distribution of the emitting chain segments accompanied by an increased effective conjugation length. It is noteworthy that this observation was made on solid nanowires, whereas for nanotubes consisting of conducting polymers obtained by polymerization in hard templates only the polymer layers in the proximity of the pore walls exhibited enhanced ordering, as discussed in Sect. 4.1 [11]. The PFO nanorods exhibited waveguide behavior, as demonstrated by local excitation of nanorod segments away from the tip and by simultaneously probing the light emission at the tip by PL microscopy. Evaluation of the intensity of the emission at the nanowire tip as a function of the propagation length revealed the occurrence of propagation losses that significantly increase with increasing distance between the excitation spot and the nanorod tip. The analysis of spatially resolved PL spectra of the waveguided emission at the nanorod tip revealed that the attenuation of shorter-wavelength peaks was stronger than that of longer-wavelength peaks, a finding that was attributed to reabsorption. However, Rayleigh scattering resulting from density fluctuations in the waveguide material on length scales one to two orders of magnitude shorter than the wavelength of the propagating light apparently contributed to the observed losses in a significant manner. The presence of such fluctuations giving rise to local variations in the refractive index along the nanorods was visualized by dark-field TEM, revealing that the nanorods contain some crystallites with diameters of a few tens of nm with orientations deviating from that of the matrix.

These results suggest that a careful and precise engineering of the crystalline morphology in one-dimensional nanostructures consisting of semicrystalline polymers is indispensable for the optimization of their performance as device components in real-life applications. Furthermore, the investigation of the crystallization in polymeric nanotubes and nanorods aligned in the nanopores of rigid hard templates is complementary to studies on the crystallization of semicrystalline blocks in microphase-separated

BCPs [117–121]. The concepts discussed above may help to understand the crystallization behavior of bulk polymers as well as that of non-polymeric materials such as pharmaceuticals [122] and inorganic semiconductors [123] confined to nanoporous hard templates.

4.3
Columnar Mesophases

Besides the exploitation of melt crystallization in thermoplastics, the formation of ordered assemblies consisting of molecules having an anisotropic shape is a self-assembly process with great potential for the fabrication of nanotubes with a customized mesoscopic fine structure inside the pores of hard templates. To this end, particularly disc-like molecules that self-organize into columnar stacks, so-called discotics [124], are promising building blocks. For their anchoring to the surface of a substrate, two limiting cases can be formulated. "Edge-on" orientation means that the molecular planes are oriented normal to the surface of the substrate. The columns formed by the disk-shaped molecules then have a so-called "planar" orientation. "Face-on" anchoring means that the molecular planes are parallel to the substrate surface. Then, the orientation of the columns is called "homeotropic" (Fig. 15). The way how the discs assemble on a surface depends on the intercolumnar interactions between adjacent discs and the interactions between the disks and the substrate.

Particularly polycyclic aromatic hydrocarbons (PAHs) [126] have attracted considerable interest, because their pyrolysis yields nanotubes whose walls consist of graphene layers. For example, Zhi et al. deposited disc-like PAHs of the hexa-*peri*-hexabenzocoronene (HBC) type into AAO hard templates from solutions in dichloromethane [103]. The discs were anchored edge-on, and columns with a planar orientation with respect to the pore walls formed, driven by strong π–π interactions between the HBC discs. Subsequent pyrolysis yielded nanotubes whose walls consisted of highly ordered graphene

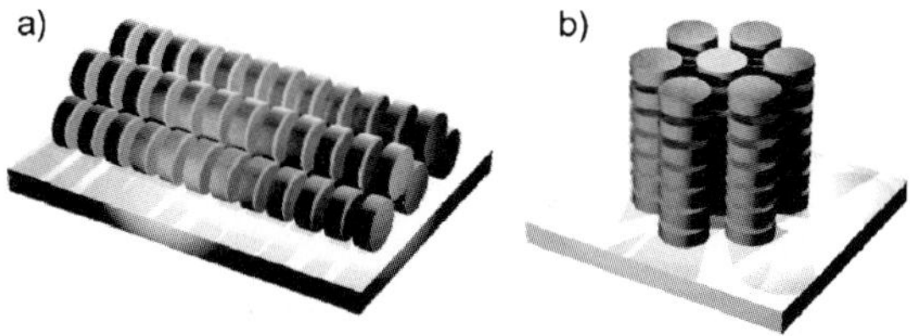

Fig. 15 Schematic representation of the different types of supramolecular arrangements of discotics on surfaces with **a** edge-on orientation of the molecules, where the columnar axis is oriented parallel to the substrate (planar texture), and **b** face-on arrangement of the discotics leading to a homeotropic texture. Reproduced from [125]. © (2005) Wiley-VCH

layers oriented perpendicular to the tube axes. Therefore, the initial orientation of the HBC precursors was conserved during the carbonization step. The properties of the HBC molecules can be engineered by modifying the substitution pattern at the polycyclic aromatic core. Long, branched alkyl side chains as substituents lead to low isotropization temperatures so that melts of correspondingly designed HBC molecules could be infiltrated into AAO hard templates [127]. In this case, the discs formed columns oriented planar with respect to the pore walls in which the plane of the disks was inclined by 45° with respect to the column axis and the long axes of the template pores. HBC molecules bearing acrylate units at the end of six alkyl spacers attached to the polycyclic aromatic core were synthesized by Kastler et al. [128]. The HBC molecules thus modified can easily be cross-linked via the acrylate functions. Deposition of these HBC discs from a solution in dichloromethane into AAO hard templates led to the formation of nanotubes whose walls consisted of long-range ordered stacks of crosslinked HBC molecules aligned with the nanotube axes. Apparently, the discs were anchored edge-on to the pore walls. Cross-linking at a moderate temperature of 170 °C fixated the supramolecular columnar architecture without destroying the polycyclic aromatic core of the molecule. The nanotubes thus obtained therefore exhibited high mechanical stability, whereas the supramolecular order, the formation of which had been driven by π–π interactions between the HBC discs, was conserved as revealed by HRTEM and SAED (Fig. 16). In the case of hyperbranched tetraphenylcyclopentadienone building blocks deposited from solutions in dichloromethane, crosslinking by a Diels–Alder reaction fixated the supramolecular architecture, and stable tubular nanostructures were obtained. Subsequent carbonization yielded carbonaceous nanotubes exhibiting a highly porous fine structure [129].

Self-organization driven by π–π interactions has meanwhile been exploited to assemble nanotubes from even more complex building blocks. To this end, Zhi et al. deposited tetrakis(tert-butyl)-naphthalocyaninato nickel complexes that form columnar structures on the pore walls of AAO hard templates by edge-on π–π stacking from a solution in THF [102]. After thermal annealing, a highly ordered columnar structure was obtained, and stable tetrakis(tert-butyl)-naphthalocyaninato nickel nanotubes could be released from AAO hard templates or converted into graphitic carbon nanotubes containing nitrogen or nickel nanoparticles. Liu et al. prepared nanotubes from sandwich-type (porphyrinato)(phthalocyaninato)europium(III) complexes [130].

However, the formation of mesophases inside the two-dimensional confinement of nanopores can be complex. For example, a model compound based on the triphenylene motif was melt-infiltrated into AAO hard templates. It was found that homeotropically anchored columns whose growth starts from the pore walls compete with planar columns in the center of the pores that proceed along the pore axes, i.e., along the direction that is free

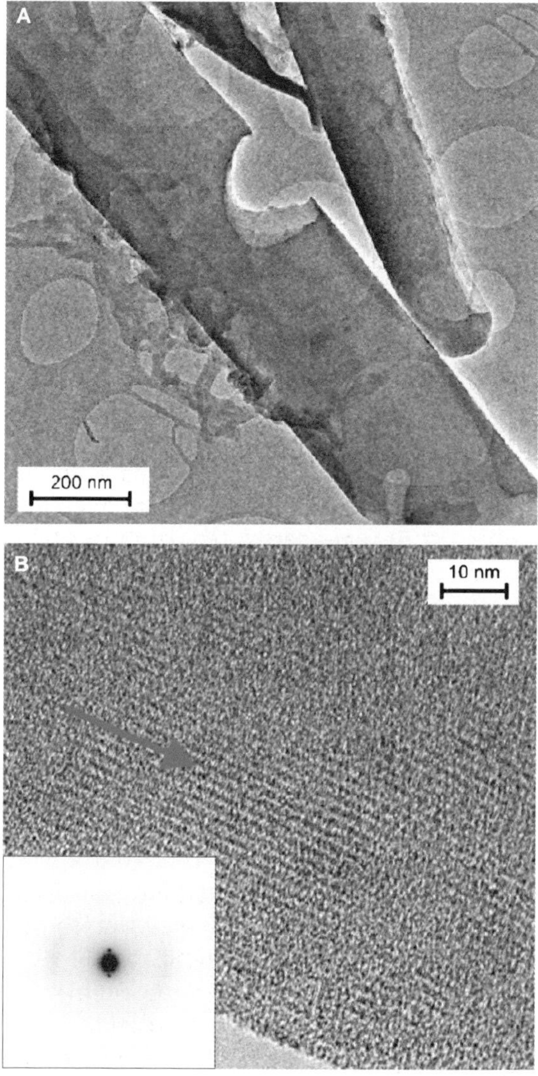

Fig. 16 TEM images of carbonaceous nanotubes obtained by assembling HBC molecules inside hard templates. **A** Defect evidencing their tubular structure. **B** Detail of the wall structure; the *arrow* indicates the directions of the columnar structures and the tube axis; *inset*: electron-diffraction pattern. Reproduced from [128]. © (2007) Wiley-VCH

of geometric constraints [65]. Whereas the planar phase was characterized by a narrow orientation distribution, the homeotropic phase was significantly less ordered. Pentacene nanotubes prepared by melt infiltration and slow cooling to room temperature [131] exhibited no long-range order. Also, HBCs that were designed in such a way that they anchor face-on on smooth substrates did not yield nanotubes with uniformly oriented graphene layers when deposited into AAO templates [132]. Thus, it seems that the suppression of face-on anchoring leading to homeotropic orientation of the columns, for example by using molecular building blocks that tend to intercolumnar π–π stacking, results in the formation of highly ordered planar mesophases. However, homeotropic phases inside AAOs are characterized by a high degree of disorder. On the one hand, the curvature of the pore walls prevents a perfect parallel arrangement of the columns along the perimeter of the pores so that growing columns will impinge on their neighbors. On the other hand, the roughness of the pore walls may also introduce disorder.

5
Phase Separation in Nanoporous Hard Templates

5.1
Spinodal Decomposition in 2D Confinement

Phase separation processes have been widely used to generate mesostructured materials, such as controlled porous glasses [133]. It is obvious that the decomposition of mixtures in the pores of hard templates is a promising access to nanotubes with tailor-made fine structures. For example, nanotubes with walls exhibiting a microporous fine structure are potential components for storage devices or chromatographic separation processes. Nevertheless, this strategy is, up to now, largely unexplored, whereas a plethora of publications deal with decomposition processes in the course of electrospinning [134–136]. Phase boundaries may be crossed if a mixture is subjected to thermal quenching, or if the composition of the mixture changes because of the evaporation of a volatile solvent. Commonly, a spinodal decomposition [137, 138] sets in. Then, periodic composition fluctuations in an initially homogeneous system begin to grow. Simultaneously, ripening of the morphology starts, driven by the tendency to minimize the interfacial area between the coexisting phases, at the initial stage by conformational changes of the polymer chains and subsequently by Ostwald ripening [139]. For a broad composition range, the phase morphology generated by spinodal decomposition is initially a bicontinuous network of the two components that breaks up at later stages of the ripening process. The presence of interfaces modifies the decomposition process in that a surface-induced layered structure forms, a phenomenon known as "surface-directed spinodal decomposition" [140]. In thin films, the concentration waves emanating from the upper and lower surfaces may interfere [141]. If the film thickness is further reduced, a crossover from three- to two-dimensional spinodal decomposition kinetics occurs [142]. Therefore, it is to be expected that decomposition in a cylindrical or tubular geometry in contact with pore walls acting as a rigid, non-critical matrix phase is modified as compared to bulk systems or thin film configurations. Predominantly liquid/liquid decomposition of binary mixtures in cylindrical pores has been studied theoretically [82, 83, 143–146]. As summarized by Gelb et al. [83] persistent metastable states occur within the two-phase region, resulting in common hysteretic behavior and a dependence of the phase morphology on the history of the sample, whereas macroscopic phase separation does not take place on experimentally accessible timescales. Moreover, critical temperatures and mole fractions are shifted.

If the pore walls are neutral with respect to the components of the mixture, a plug-like morphology develops with a domain size that saturates at a length scale far away from macroscopic phase separation [143, 144]. However, particularly the affinities of the components to the pore walls are important parameters governing decomposition processes in nanopores. Phase separation induced by thermal quenching might be accompanied by wetting transitions, that is, changes in the relative affinities of the components to the pore walls [84]. Dependent on the temperature, two equilibrium states were predicted. In the case of partial wetting, where both components are in contact with the pore walls, the equilibrium morphology consists of two cylindrical segments consisting of the pure components, which are separated by one interface stretching across the pore. However, long-lived metastable states characterized by a sequence of short plugs, each of which is in contact with the pore walls, were predicted to be more realistic. If the pore walls are completely wetted by one of the components, the non-wetting component forms a cylinder in the center of the pore surrounded by the wetting component that is exclusively in contact with the pore walls.

Apparently, no experimental studies of liquid/liquid decompositions of binary polymer mixtures confined to nanopores have been reported up to now. A possible reason for this lack of experimental work lies in the fact that the in situ monitoring of phase separation inside nanoporous hard templates is far from being trivial. As discussed above, the background scattering of the hard template is a severe drawback for SAXS. The information accessible by other methods such as DSC is limited. Moreover, given the importance of interfacial effects, it appears to be difficult to infiltrate molten polymer blends in such a way that the composition inside the pores corresponds to that of the bulk mixture on top of the hard template. It is reasonable to assume that the component with lower affinity to the pore walls will be expelled from the pores. The homogeneity of polymer blends on the mesoscopic length scales relevant to the infiltration of nanopores is certainly another issue to be considered. The most straightforward way to infiltrate mixtures containing at least one polymeric component into nanoporous hard templates is wetting with homogeneous solutions in a common volatile solvent. However, the presence of a third, evaporating component complicates the understanding of the involved structure formation processes. The analysis of the phase morphology obtained in this way was, up to now, limited to SEM and TEM investigations of nanotubes and nanorods in which the structure evolution was frozen.

As model systems, mixtures of PLA and organometallic complexes containing Pd and Pt dissolved in a common solvent were filled into the pores of hard templates [85, 147]. After the evaporation of the solvent, tubes with a wall thickness of a few tens of nm were formed in which the metal precursors were dispersed. After thermolytic reduction, the evolution of nanoparticles consisting of the elemental metals inside the liquid polymeric matrix

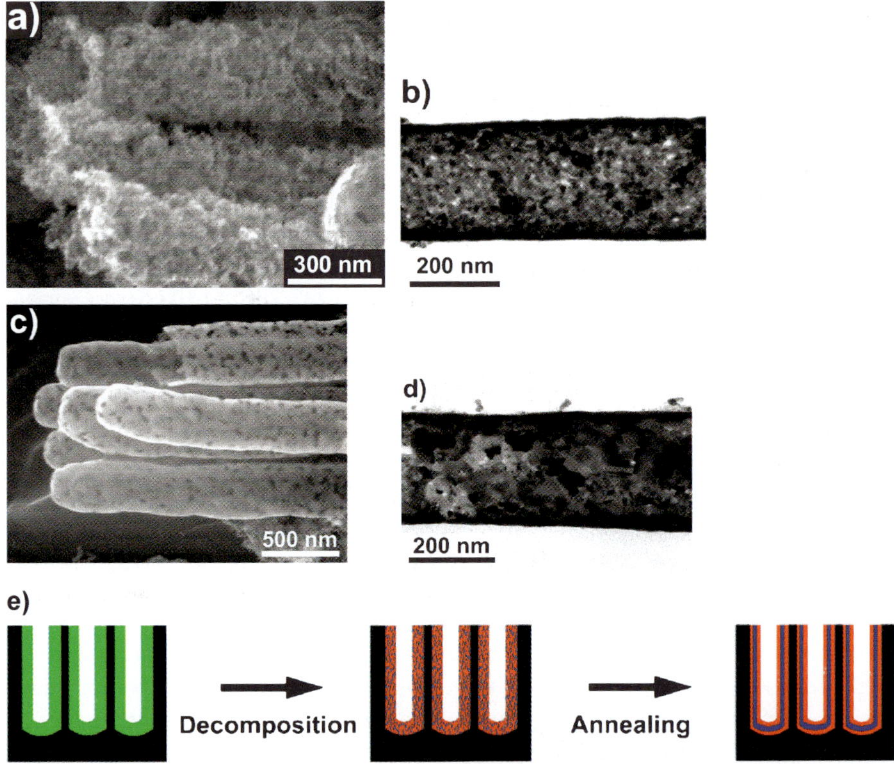

Fig. 17 Palladium nanotubes obtained by the ripening of Pd nanoparticles in a PLA matrix after spinodal decomposition of a PLA/palladium acetate/solvent mixture inside AAO and pyrolytic degradation of PLA. **a** SEM and **b** TEM image after short ripening times; **c** SEM and **d** TEM image after longer ripening times. **e** Schematic diagram of the ripening process. After the decomposition of the initially homogeneous mixture, the walls of the Pd nanotubes were at first rough, highly porous, and had obviously a reticular structure, which is indicative of an interpenetrating morphology with a small spinodal wavelength. At later ripening stages, the nanotube walls had a smoother, layer-like appearance, and the size of the Pd crystals significantly increased, which is indicative of the evolution of a coarser, layered structure and surface-induced ordering. Panels **a–d** are reproduced from [63]. © panels **a–d** (2004) Wiley-VCH

was monitored as a model process for spinodal decomposition and morphology ripening. The ripening was stopped by pyrolytic degradation of the PLA. After short ripening times, for example, the walls of Pd nanotubes were rough, highly porous, and had obviously a reticular morphology (Fig. 17a,b). At later ripening stages, the nanotube walls had a smoother, layer-like appearance, and the size of the Pd crystals significantly increased (Fig. 17c,d). Thus, after short ripening times, the initially homogeneous tube walls are characterized by an interpenetrating morphology with a small spinodal wavelength. Further ripening results in the evolution of a coarser, layered structure that is indicative of surface-induced ordering (Fig. 17e). Wetting AAO templates with solutions containing a polymeric wetting carrier and precursors for magnetic metals such as cobalt was applied to synthesize magnetic nanotubes [148, 149].

Surface-induced ordering was also observed in nanotubes obtained by deposition of a solution containing PMMA and a discotic liquid crystal of the triphenylene type into AAO templates with a D_p-value of 400 nm. The PMMA segregated to the pore walls, whereas the liquid crystal enriched at the inner surface of the nanotubes (Fig. 18a). It was assumed that the synergistic interplay of two different physico-chemical phenomena led to a surface-directed phase separation. First, low molecular mass species such as the triphenylene compound enrich at interfaces in the presence of a polymer for entropic reasons [150, 151]. Secondly, taking into account the high compatibility of the liquid crystal used and PS, PMMA should have a higher affinity to the AAO pore walls [152–154]. Reducing the D_p-value of the AAO hard template to 60 nm resulted in the occurrence of a striking morphological crossover. Solid nanorods were obtained with a disordered segmented morphology (Fig. 18b). The disappearance of the inner tube surface obviously resulted in a competition of the enthalpic and entropic effects: the PMMA tends to segregate to the pore walls for enthalpic, and the triphenylene compound for entropic rea-

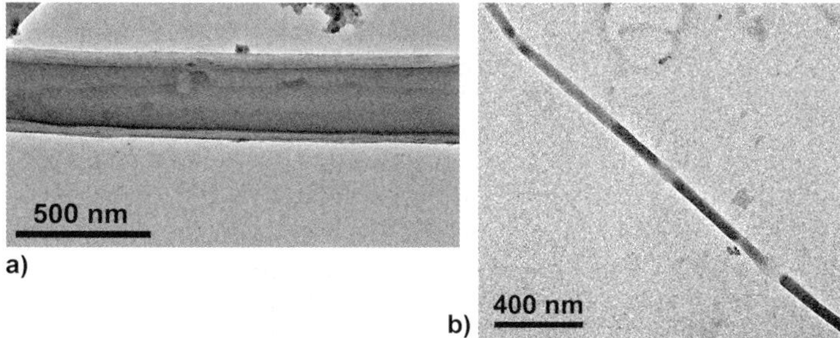

Fig. 18 PMMA/discotic composite nanofibers prepared inside AAO hard templates. **a** Nanotube (diameter about 400 nm) with a wall consisting of an outer PMMA layer and an inner (stained) discotic layer; **b** nanorod (diameter 60 nm) with a disordered segmented morphology. Reproduced from [65]. © (2005) Wiley-VCH

sons. Since only one interface at the pore walls instead of two interfaces in a tubular configuration was available, the formation of a layered structure was prevented. Thus, a confinement-induced transition from a wetting state (one phase of a critical two-phase system in exclusive contact with an interface) to a non-wetting state (both phases in contact with the interface) occurred.

5.2
Sol/Gel Chemistry with Block Copolymer Soft Templates

Spinodal decomposition is a straightforward access to mesostructured materials characterized by a certain degree of near order. However, phase separation can even be exploited to fabricate ordered mesostructured materials if amphiphilic species acting as soft templates are involved. Well-established syntheses for mesoporous materials with mesopore diameters ranging from a few nanometers up to a few tens of nanometers start with sol solutions containing either low molecular mass surfactants [155–157] or BCPs containing blocks with different polarity [158, 159] and precursors for scaffolds consisting of inorganic oxides or amorphous carbon [160–165]. The precursor for the scaffold material commonly segregates into the polar phase defined by the soft template that self-assembles if its concentration is larger than the critical micelle concentration. Subsequently, the morphology thus formed is fixated by a gelation or aging step in which the precursors for the inorganic scaffold materials are crosslinked. Subsequent high-temperature calcination yields inorganic scaffolds containing highly ordered mesopore arrays. Several excellent review articles summarize syntheses for mesoporous materials and their properties [166–172]. However, they are typically obtained in the form of powders consisting of randomly oriented grains. On solid substrates, the mesopores are arranged parallel to the substrate surfaces, a morphology that is of limited use for applications in the fields of separation, catalysis, and storage. Strategies to overcome this drawback based on surface modifications or the freezing of non-equilibrium structures only yield thin mesoporous layers with mesopores having limited aspect ratios. Therefore, it is still challenging to fabricate free-standing mesoporous membranes having the mesopores oriented normal to the plane of the membrane by approaches based on sol/gel chemistry. In order to address this problem, the self-assembly of BCP soft templates inside the pores of hard templates has emerged as a promising synthetic strategy, taking advantage of the availability of mechanically stable, extended membranes having larger but properly oriented membrane pores (Sect. 2).

The first reported procedures for the preparation of mesoporous silica nanofibers by means of BCP soft templates inside hard templates were adaptations of synthetic strategies introduced by Stucky and coworkers [159]. Both tubular and solid entities with a mesoporous fine structure can thus be prepared. In their pioneering work, Yang et al. filled sols containing Pluronic

F127 and TEOS into AAO hard templates with a D_p-value of about 250 nm. Inside non-modified AAO the mesoporous silica remained attached to the oxidic, polar pore walls of the hard template so that tubes formed. However, if the hard templates were modified with hydrophobic silane coupling agents, the silica detached from the pore walls of the hard template, and solid rods were obtained [173]. Liang and Susha reported that infiltration of sols into polycarbonate membranes yielded tubular structures if the sols had high ethanol content, whereas solid rods formed when the ethanol content was reduced [174]. Both Liang and Susha, as well as Yao et al. [175] moreover found that aging the sols inside the pores of a hard template in the presence of an external bulk sol reservoir promotes the formation of solid mesoporous silica fibers, whereas the removal of excess sol from the surfaces of the hard templates after the infiltration promotes the formation of tubular structures. Zhu et al. obtained tubular structures in AAO templates with a D_p value of about 200 nm by slow infiltration of mixtures containing prehydrolized TEOS [176]. To this end, a sol solution was cast onto a smooth substrate. After placing AAO hard templates on top of the sol films, the samples were annealed at 100 °C. The lengths of the silica nanotubes thus obtained ranged from 500 nm for a heating time of 2 h up to 10 μm for longer heating times. Apparently, the prehydrolized sol could slowly enter the pores governed by the capillary wetting mechanism (Sect. 3.3) but the reduced mobility of the sol apparently prevented the filling of the empty space inside the pores formed upon evaporation of the solvent. Thus, it appears that the nature of the pore walls, the composition of the infiltrated sol, and the amount of sol solution that can access the pores are parameters determining whether tubular or solid mesoporous silica structures form inside hard templates.

Gaining control over the orientation of the mesopores inside hard templates has turned out to be a delicate challenge. Subtle changes of parameters such as the composition of the sol or aging conditions lead to striking changes of the mesopore structure, which is in turn affected by the two-dimensional geometric confinement imposed by the pore geometry and the nature of the pore walls. In the case of mesoporous materials, the bulk morphology of which is characterized by hexagonal arrays of aligned mesochannels, the mesopores formed by self-assembly of BCP soft templates inside the pores of hard templates may align with the long axes of the hard template pores, or they adopt the contour of the pore walls and wind about the long axes of the hard template pores. If the hard templates have pores smaller than about 100 nm, new and unprecedented morphologies are obtained that are substantially different from their bulk counterparts, as discussed below. Self-assembly of different BCP soft templates of the Pluronic type inside the pores of hard templates with D_p-values of a few 100 nm yielded hexagonal arrays of mesopores winding about the long axes of the mesoporous silica nanofibers [173, 174, 177] (Fig. 19), as well as hexagonal mesochannel ar-

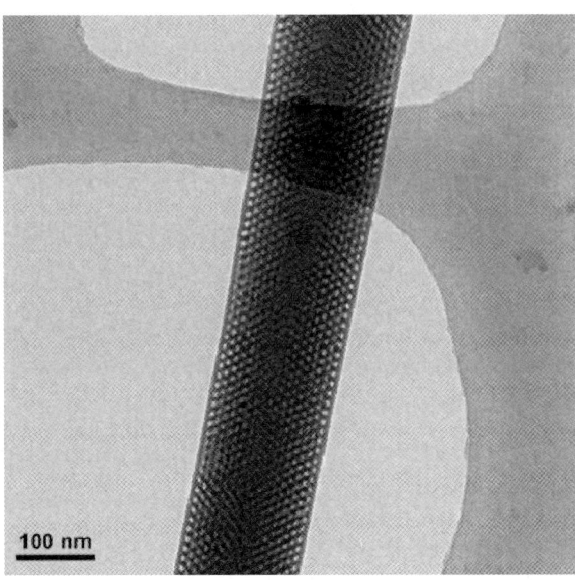

Fig. 19 Example of a mesoporous silica nanofiber containing hexagonal arrays of mesopores winding about the fiber axis. Reproduced from [177]. © (2005) American Chemical Society

rays [178] and concentric-lamellar structures [179] aligned with the fiber axes.

The relative orientation of hexagonal mesopores inside hard templates with a D_p-value of about 200 nm was studied by several authors and depends both on the composition of the sol and on the aging conditions. Yang et al. reported that an increase in the concentration of the BCP soft template in the sol solution led to a crossover from circular mesopores perpendicular to the long axes of the hard template pores to mesoporous channels aligned with the long axes of the hard template pores [173]. Platschek et al., who investigated the self-assembly of BCP and surfactant soft templates in AAO hard templates by means of TEM and grazing incidence SAXS, reported that a drastic shift in population from a circular pore arrangement towards a columnar arrangement of the mesopores aligned with the long axes of hard template pores (Fig. 20) occurred along with an increase in surfactant concentration in the sols or environmental humidity [180]. Under conditions characterized by high humidity and high surfactant concentration, the time required for evaporation of the solvent was extended so that a columnar morphology appearing to be closer to equilibrium could be attained. The presence of water during gelation was also highlighted as a key parameter determining the pore arrangement by other authors. Jin et al. observed mesopores partially aligned with the hard template pore axes in the center of the mesoporous entities surrounded by circular mesopores in samples aged in the absence of

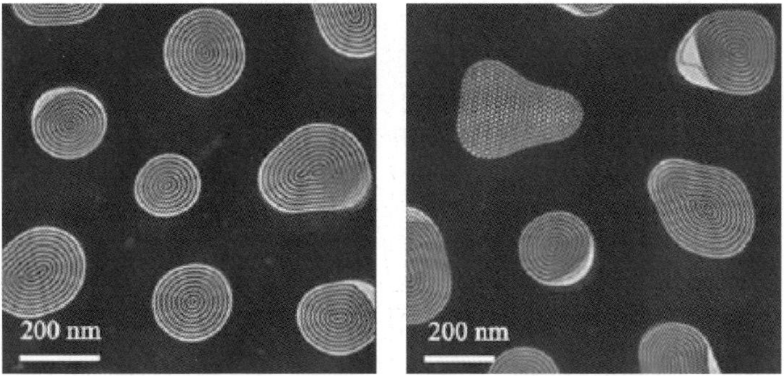

Fig. 20 Plan-view TEM images of mesoporous silica nanofibers with dominant circular orientation of the mesopores. Reproduced from [180]. © (2006) Wiley-VCH

water, whereas in samples aged in the presence of water a circular arrangement of the mesopores was found [181]. These authors assumed that the presence of water in the sol would retard the evaporation of the volatile solvents and concluded that the circular pore configuration would correspond to the equilibrium. Yao et al. reported that in as-prepared silica nanofibers the mesochannels are oriented parallel to the long axes of the hard template pores in the absence of water, or perpendicular in the presence of water during aging and related this observation to the acceleration of the hydrolysis of the silica precursor TEOS caused by the presence of water [175]. Hence, the configuration obtained in the absence of water, that is, hexagonal mesopore arrays parallel to the long axis of the hard template pores would correspond to the equilibrium. Despite these somewhat inconsistent findings and conclusions drawn by different authors, it seems to be unambiguous that the stage at which the assembly of the soft template is frozen determines the morphology of the mesoporous nanofibers. Apparently, the mesoscopic structure results from counteracting growth modes governed by surface-induced ordering and the two-dimensional confinement imposed by the geometry of the hard template pores. It is noteworthy that inside macroporous silicon with a pore diameter of 1 μm no preferred mesopore orientation was found. Instead, the mesoporous microfibers consisted of segments with different mesopore orientations [19].

An interesting procedure for the incorporation of cadmium selenide nanoparticles into mesoporous silica nanofibers involves the use of sols containing Cd(II) species [177]. After the gelation step and prior to calcination, the Cd was converted into CdS by exposing the samples to gaseous H_2S. Whereas in the case of a small Cd(II) content in the sol solutions hexagonally ordered circular mesopores winding about the long axes of the hard template pores were obtained, higher Cd(II) contents led to the generation of disordered pore structures along with a significant increase in the specific surface, as determined from adsorption isotherms. Also, the average size of the CdSe nanoparticles could be increased by increasing the Cd(II) content in the sol, as evidenced by absorption spectroscopy and the shift of the band-edge emission in the PL spectra.

Whereas in the works reviewed above predominantly AAO hard templates with a D_p-value about one order of magnitude larger than the periods of the BCP soft templates were used, Wu et al. systematically studied the self-assembly of $EO_{20}PO_{70}EO_{20}$ inside AAO hard templates having D_p-values from 80 down to 20 nm [182, 183]. Inside such narrow pores, bulk-like morphologies were completely suppressed. For D_p-values between 55 and 73 nm,

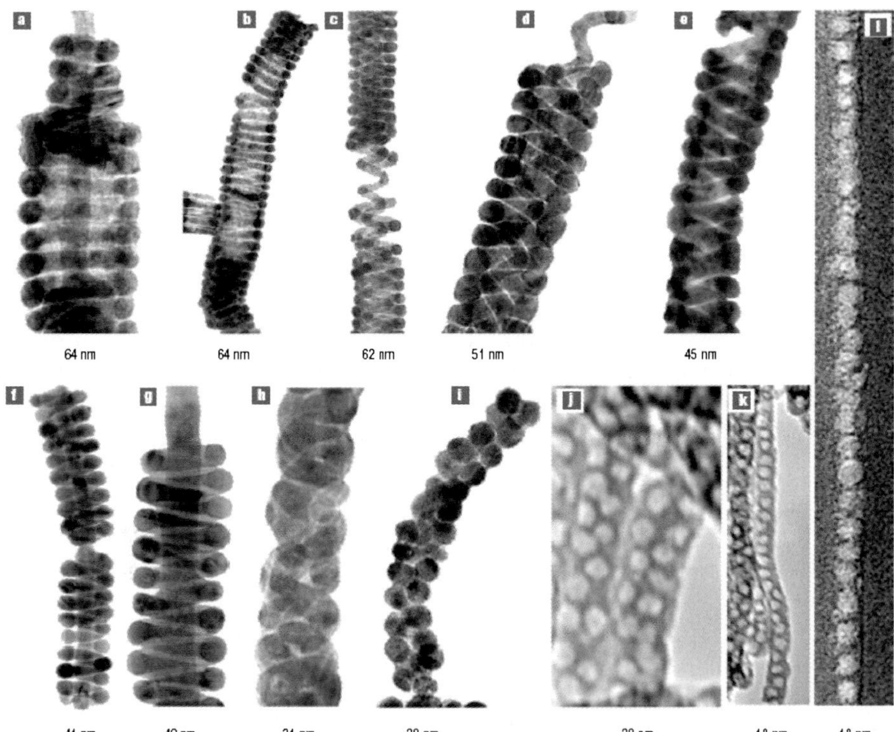

Fig. 21 TEM images of mesostructures formed inside AAO with differing confinement dimensions. The confining nanochannel diameter is indicated underneath each image. **a–i** Silver inverted mesostructures prepared by backfilling the confined mesoporous silica; **j–k** free-standing mesoporous silica fibers; **l** mesoporous silica embedded inside the AAO obtained using a focused ion beam for sample preparation. The structures are **a** three-layer stacked doughnuts; **b** S-helix; **c** core–shell D-helix, in which the core and the shell are both S-helix; **d** core–shell triple-helix, in which the shell is a D-helix and the core is a S-helix; **e** D-helix; **f, g** S-helix with a straight core channel; **h** D-helix; **i, j** inverted peapod structure with two lines of spherical cages packed along the long axis of the alumina nanochannel; **k, l** inverted peapod with one line of cages. Reproduced from [182]. © (2004) Nature Publishing Group

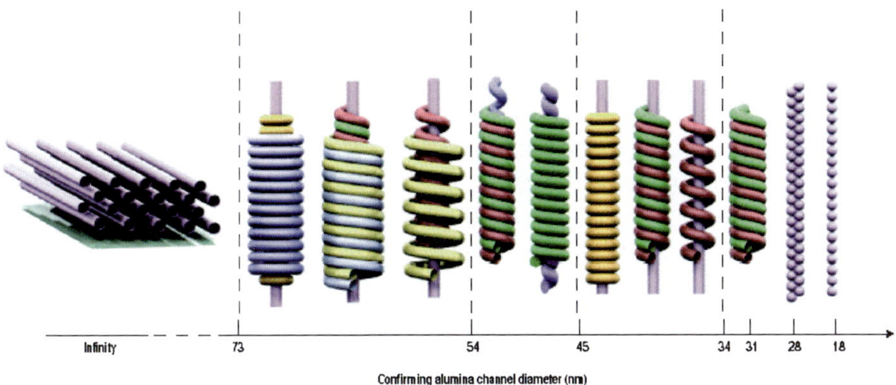

Fig. 22 Summary of the experimentally (cf. Fig. 21) observed confined mesostructural evolution with varying D_p-value. Reproduced from [182]. © (2004) Nature Publishing Group

the mesostructures were composed of a straight core and two more coaxial layers consisting of concentric mesochannels with morphologies as diverse as stacked doughnuts, single helices or double helices (Figs. 21, 22). For D_p-values ranging from 49 to 54 nm, coaxial double layer helices were found, for D_p-values ranging from 34 to 45 nm a straight inner core was surrounded by one coaxial layer of the helical or stacked doughnut type, for a D_p-value of 31 nm a single helix was observed, and for smaller D_p-values spherical mesopores arranged in one or two rows formed (Figs. 21, 22). Analogous to the conceptualization of the structures of carbon nanotubes, Wu et al. suggested a rolling scheme to derive the morphologies obtained in the two-dimensional confinement of the hard template pores from thin-film morphologies. A progression of mesoscopic structures in two-dimensional confinement well in line with the experimental results was obtained by means of self-consistent field calculations with a liquid diblock copolymer/homopolymer mixture as a model system. Moreover, it is remarkable that chiral structures such as helices were obtained from achiral materials, even though a chiral induction leading to enantiomeric excess has not been reported up to now.

One of the few reports on the preparation of mesoporous silica nanofibers with a soft template other than triblock copolymers of the Pluronic type deals with the synthesis of silica nanofibers with high aspect ratios containing linear arrays of mesopores by a solution-induced self-assembly process (Fig. 23), as previously reported for thin-film configurations [184]. To this end, PS-*b*-PEO diblock copolymers were employed as structure-directing agents in sol solutions containing toluene/ethanol mixtures. For a D_p-value of 35 nm, a single line of mesopores formed, for a D_p-value of 60 nm two parallel rows of mesopores were obtained [185].

Fig. 23 TEM image of released silica nanowires containing a single row of mesocages obtained with PS(9500)-*b*-PEO(9500) as the soft template and an AAO hard template (D_p = 35 nm). Reproduced from [185]. © (2007) Wiley-VCH

Few efforts have been directed towards the fabrication of mesoporous nanofibers consisting of other inorganic oxides. Mesoporous titania nanofibers containing anastase crystallites with diameters of the order of 6.5 nm were obtained by Chae et al. by infiltration of sols containing $EO_{106}PO_{70}EO_{106}$ as a soft templat and titanium(IV)isopropoxide as a titania source [186] into AAO hard templates with a D_p-value of 200 nm. However, the mesopore arrays did not show long-range order. In a modified procedure with a low-viscous sol solution, Wang et al. fabricated nanotubes the walls of which consisted of mesoporous titania exhibiting hexagonal mesopore ordering [187]. The evolution of a tubular structure was attributed to the high affinity of the gel to the alumina pore walls, resulting in volume shrinkage towards the pore walls of the hard template.

As compared to mesoporous oxide nanofibers, much lesser attention has been paid to their mesoporous amorphous carbon analogues. However, mesoporous carbon exhibits superior resistance to acids and bases, excellent heat resistance, as well as high intrinsic electric conductivity. Potential applications for hybrid membranes consisting of mesoporous carbon within hard templates include size-selective electrosorption, electrosynthesis of nanostructures, catalysis, separation and storage. The first reported procedure for the synthesis of mesoporous carbon nanofibers involved the preparation of a sacrificial Fe-containing mesoporous silica scaffold inside an AAO hard template with a D_p-value of about 200 nm, exposure to hydrogen at 750 °C, incorporation of carbon by supercritical fluid deposition of a xylene/CO_2 mixture, and removal of both the AAO hard template and the sacrificial silica scaffold by etching with hydrofluoric acid [188]. While the mesoscopic fine structure of the mesoporous carbon nanofibers was a perfect replica of the silica scaffold, the AAO hard template is inevitably destroyed upon removal of the silica scaffold. Consequently, hybrid membranes containing mesoporous carbon inside an AAO matrix are not accessible by this approach. This drawback can be overcome by directly synthesizing mesoporous amorphous carbon inside the pores of the hard template. Zheng et al. infiltrated a mixture of Pluronic F127 ($EO_{106}PO_{70}EO_{106}$) as a structure directing soft template and resol as a carbon precursor dissolved in ethanol into AAO hard templates. After the evaporation of the ethanol, gelation and carbonization at 700 °C in nitrogen, mesoporous carbon nanowires were obtained. Inside AAO hard templates with a D_p-value of about 300 nm a core/shell structure was obtained in which a stack of layers perpendicular to the nanowire axis surrounded a core containing pores winding about the nanowire axis. At the same time, an approach based on solvent-free infiltration was reported. A solution of Pluronic F127, phloroglucinol as a carbon source, formaldehyde and traces of HCl in an ethanol/water mixture was stirred at room temperature until a separation into an upper water/ethanol phase and a lower polymer-rich phase occurred. The supernatant solvent-rich phase was removed, and the lower polymer-rich phase was spread on AAO hard templates. Gelation of the infiltrated mixture and subsequent calcination at 500 °C yielded mesoporous amorphous nanofibers with a core characterized by a bicontinuous morphology, as desired for applications in the field of separation, catalysis and storage [189]. Again, a layered shell indicative of surface-induced ordering was found that was in turn surrounded by a continuous outermost carbon wall. Whereas the removal of volatile solvents prior to the infiltration may be a measure to minimize volume shrinkage, the low carbonization temperature is important for the fabrication of mesoporous amorphous carbon/AAO hybrid membranes, because the carbonization can be performed while the AAO membrane is still attached to an underlying Al substrate. This configuration is advantageous because the Al substrate stabilizes the AAO layers so that residual material can easily be removed from their surfaces to uncover the pore openings. Selective etching steps can then be applied to remove the Al and to open the pore bottoms. Optionally, an Al ring surrounding the area in which the pore bottoms are open can be conserved to mechanically stabilize the membrane. It is interesting to note that the fiber core vanishes when the D_p-value of the hard template is decreased below 100 nm. Arrays of freestanding mesoporous amorphous carbon nanofibers on silicon substrates using Pluronic F-127 as a structure-directing soft template were prepared by Wang et al. by placing an AAO hard template on the substrate [190]. After gelation and calcination, a supercritical drying process was applied to avoid condensation of the aligned mesoporous amorphous carbon fibers after the wet-chemical etching of the hard template. Their internal mesopore morphology resembled that previously found for mesoporous silica nanofibers and consisted of circular mesopores perpendicular to the long axes of the nanofibers that occasionally surrounded a core of mesopores aligned with the long axis of the nanofibers.

It is obvious that hybrid systems of AAO hard templates containing mesoporous nanofibers obtained by self-assembling BCP soft templates are of considerable interest for a plethora of applications in the fields of catalysis, separation and storage. However, the combination of sol/gel chemistry and high-temperature calcination steps is accompanied by pronounced volume shrinkage of the mesoporous material. Solvent evaporation during the initial gelation step performed at room temperature results in unidirectional shrinkage of up to 20% [191], and further cross-linking as well as calcination lead to unidirectional shrinkage of about 15–40%, depending on the protocol applied [192]. Yao et al. noted that shrinkage inside AAO hard templates occurs primarily in a direction perpendicular to the AAO pores with void space being created at the silica–alumina interface [175]. Given the high affinity of the sol-gel to the pore wall, it is reasonable to assume that large-scale shrinkage in the vertical direction is prevented rather than shrinkage in the transversal direction. Consequently, the length of the hard template pore segments filled with the mesoporous material can hardly be controlled in a rational manner. Moreover, the mesoporous material occupies only a certain portion of the cross-sectional area of the hard template pores so that large voids and empty spaces occur. It is obvious that these structural defects will deteriorate the performance of the hybrid membranes.

Yamaguchi et al., who prepared microporous silica inside AAO hard templates with a D_p-value of 200 nm using cetyltrimethylammoniumbromide as a low molecular mass soft template, demonstrated size-selective separation of a set of model compounds [193]. However, to this end, a non-calcinated AAO membrane containing a silica-surfactant nanocomposite was used. Replacing calcination by extraction is another approach to overcome undesired volume shrinkage of mesoporous nanofibers inside hard templates. Yoo et al. prepared at least partially cubic mesoporous silica inside AAO with a D_p-value of about 200 nm using Brij-56 as a surfactant and removed the latter by extraction with ethanol [194]. Four successive infiltration/drying/surfactant extraction cycles yielded AAO/mesoporous silica hybrid membranes that exhibited excellent helium/nitrogen permselectivity with permselectivity values at the theoretical Knudsen limit. This result indicated the absence of "pinhole defects". Therefore, a defect-free, uniform filling of the AAO pores with mesoporous silica could be evidenced. Another interesting application reported for released mesoporous titania nanotubes, thus insensitive to volume shrinkage, is their use as nanostructured electrode material having outstandingly high specific surface area [187]. Electrons injected into the titania scaffold can rapidly be transferred into electrolyte solutions. The mesoporous titania nanotubes were characterized by large specific capacity and a high charge-discharge rate. Even though a small number of applications for either re-leased mesoporous nanofibers or hybrid membranes containing mesoporous nanofibers synthesized by means of sol-gel chemistry with BCP soft templates have been reported, it appears that these materials still need to be optimized for real-life applications.

5.3
Microphase-Separated Block Copolymer Melts

Liquid block copolymers consisting of immiscible blocks are known to self-assemble into ordered periodic arrays of nanoscopic domains [195–197]. Hence, infiltration of BCP melts in hard templates is an alternative to sol/gel chemistry that yields complex polymeric nanofibers with an adjustable fine structure. As discussed in Sect. 3.3, capillary wetting governs the filling of nanopores with BCP melts. Threads of liquid but microphase-separated BCPs preceded by menisci slowly move into the pores of the hard templates. The time scale on which this process takes place allows controlling the length of the BCP nanofibers by quenching the BCP to a temperature at which it is solid. It is to be expected that the two-dimensional geometric confine-ment and the influence of the pore walls modify the microphase structure of the BCPs, as is the case in thin-film configurations, in which a one-dimensional confinement is imposed. Structure formation processes in thin BCP films have been studied intensively [198, 199]. A scenario in which BCPs self-assemble between smooth, non-competing surfaces exhibits, to a cer-tain extent, similarities to the self-ordering of BCPs in nanopores. Whereas a large number of experimental publications deal with sol-gel chemistry in-side nanoporous hard templates, no simulations of the related structure for-mation processes have been performed so far since the system appears to be too complex. In contrast to the lack of theoretical studies of the mesophase formation in confined multicomponent sol systems, the self assembly of BCP melts under cylindrical confinement was simulated in a considerable number of works.

In the case of diblock copolymer melts, which are the simplest model system for the elucidation of structure formation processes involving BCPs in nanoporous hard templates, only the two immiscible blocks have to be considered as components. Self-consistent field methods were applied to study the morphologies of liquid diblock copolymer/homopolymer mixtures that were considered as a model system for triblock copolymers in sol so-lutions [182], of pure diblock copolymer melts [200–202], and of order-disorder transitions in diblock copolymer melts [203]. For example, Li et al. found for a model diblock copolymer that forms cylinders in the bulk a variety of novel microdomain structures. With increasing D_p-value a sequence of morphologies containing single cylinders, stacked disks, single helices, double helices, toroid-spheres, and helices-cylinders occurred [201]. Other studies dealing with the morphologies of diblock copolymers employed dynamic density functional theory [204] and dissipative particles dynamics [205, 206]. However, in most works Monte Carlo simulations were used to investigate the mesoscopic structures of BCP-containing systems. Systems that were studied include melts of symmetric [207–212] and asymmetric diblock copolymers [211–213], of triblock copolymers [214, 215], and even mixtures of diblock copolymers [216]. Taking into account that the pores in hard templates often exhibit deviations from an ideal circular cross-section, the results of Monte Carlo simulations in which channel-like confining geometries with non-circular cross-sections were assumed are of considerable interest [217].

It is beyond the scope of this review to discuss the results of the simulations available so far in detail. Instead, some general tendencies are highlighted. Four parameters that mainly govern the morphology formation of a BCP-containing system in a nanopore have been identified.

(i) The type of equilibrium morphology into which a bulk BCP with given segment-segment interactions between its constituents and a given degree of polymerization self-assembles, depends on the volume fractions of its constituents. Under cylindrical confinement, the composition plays an important role too but does not exclusively determine which kind of equilibrium structure the BCP adopts.

(ii) It is straightforward to combine the parameters that determine the degree of geometric confinement imposed on the system, that is, the overall degree of polymerization of the BCP and the D_p-value of the cylindrical pore. Often, a reduced pore diameter is introduced. This quantity was, for example, defined as D_p/L_0, where L_0 is the bulk period of the model BCP [213]. Erukhimovich and Johner, who attempted to put confined structure formation of BCPs into a more general context, suggested a dimensionless parameter that is the product of D_p and the wave number of fluctuations of a critical order parameter that determines the ordering in the corresponding bulk system [218]. Whereas the critical order parameter in the case of an order–disorder transition in a BCP melt is related to the concentration of the repeat units of its constituents, the more general definition of a reduced pore diameter allows the treatment of ordering in cylindrically confined systems to be extended to a broader range of structure formation processes.

(iii) The degree of repulsion between the blocks of the BCP has turned out to be a crucial parameter for the structure formation under cylindrical confinement, being of considerably greater importance than for structure formation in bulk systems. For example, Feng and Ruckenstein found structural transitions when the interaction parameter between segments of the two blocks was changed [211].

(iv) The "surface field" that describes surface-induced ordering is of similar importance for the structure formation of BCPs under cylindrical confinement as for corresponding thin-film configurations. Taking into account the well-known strong influence of surfaces on spinodal phase-separation processes in cylindrical pores (Sect. 5.1), it is not surprising that the affinity of the components of the BCP to the pore walls can even govern the evolution of the mesoscopic morphology. The transition from a non-neutral pore wall exclusively wetted by one of the components to a neutral pore wall that is in contact with both components can be accompanied by a fundamental morphology transition in the whole system. For example, Chen et al. reported for a symmetric diblock copolymer a transition from "stacked disk" morphology to "cylindrical barrel" morphology. In the former case, a weak surface field prevented preferential wetting, and both components were in contact with the pore walls. In the latter case, a strong surface field led to preferential wetting of the pore walls by one of the components. For weak and intermediate strengths of the surface field also helical and even catenoid-cylindrical structures occurred [210].

A plethora of new morphology types, some of which are characterized by very complex domain structures, were predicted (Fig. 24). In most publications, some tens of different morphologies identified by simulations were displayed as snapshots but not related to a quantitative order parameter [200–202, 213]. Simple morphologies found for symmetric diblock copolymers include the stacked-disc structure with lamellae oriented perpendicular to the long axes of the hard template pores, which occurs for weak surface fields, and a concentric cylinder structure with circular lamellae oriented parallel to the long axes of the hard template pores in the case of strong surface fields. Lamellar structures of the minority component in asymmetric block copolymers can be induced by the surface field emanating from the pore walls [201], and the occurrence of helical structure motifs was predicted for a broad range of diblock copolymer compositions, predominantly in the presence of weak surface fields. It should also be noted that the mesopore structures of the silica nanofibers obtained by Wu et al. by self-assembling a $EO_{20}PO_{70}EO_{20}$ triblock copolymer in a sol solution infiltrated in AAO membranes [182] fit into the overall picture. It is again remarkable that confinement-induced chiral structures are generated in systems containing only achiral components. As pointed out by Erukhimovich and Johner [218], the free energies of both enantiomeric morphologies are identical so that statistically equally frequent formation of right-handed and left-handed helices is to be expected.

The phase behavior of asymmetric diblock copolymers seems to be more complex than that of symmetric diblock copolymers and the phase behav-

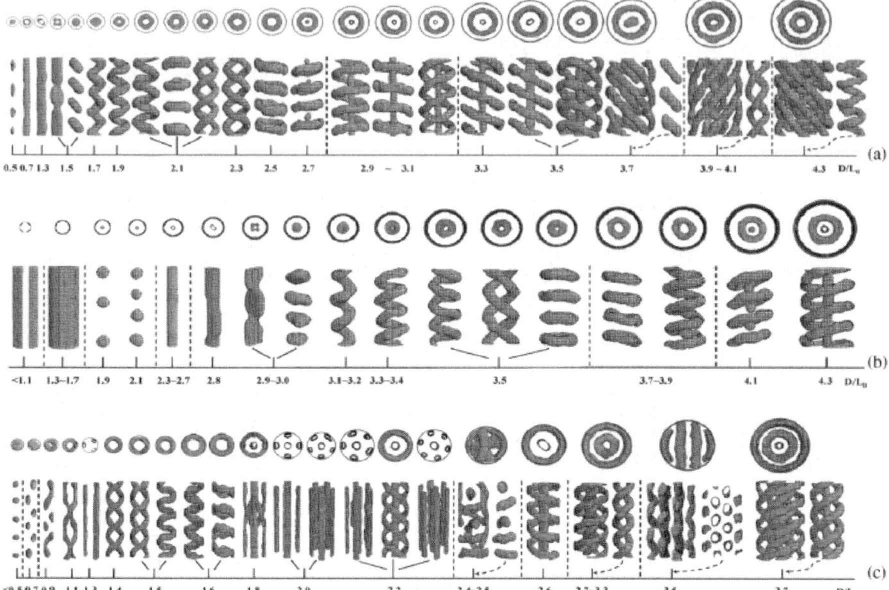

Fig. 24 Self-assembled morphologies of an asymmetric diblock copolymer confined to cylindrical pores obtained by a simulated annealing method as a function of the ratio D_p/L_0, where L_0 is the period of the BCP, for different wall-polymer interactions. The parameter D_p/L_0 is given underneath each morphology. The outmost circles in the top views indicate the wall of the cylindrical pores. For some large diameters, the inner ring is shown separately. **a** The pore wall attracts the majority blocks; **b** the pore wall attracts the minority blocks; **c** neutral pore walls. Reproduced from [213]. © (2006) American Physical Society

ior of triblock copolymers is more complex than that of diblock copolymers. In principle, it should be possible to determine phase diagrams displaying the dependence of the morphology type into which BCPs self-assemble under cylindrical confinement on each of the four parameters mentioned above, i.e., the relative volume fractions of the blocks, the reduced pore diameter, the segmental interaction parameters of the components and the surface field. However, apart from visualizing morphology snapshots, only a few attempts have been made to develop classical phase diagrams based on the relation of the body of identified morphologies to a quantitative order parameter. Erukhimovich and Johner derived reduced phase diagrams in which, for example, phase boundaries were displayed in the reduced temperature-reduced D_p plane for different surface field strengths [218]. Here, the reduced D_p-value is, as mentioned above, the product of D_p and the wave number of fluctuations of a critical order parameter determining the ordering in the corresponding bulk system, and the reduced temperature is related to the bulk order–disorder transition. Li et al. presented a phase

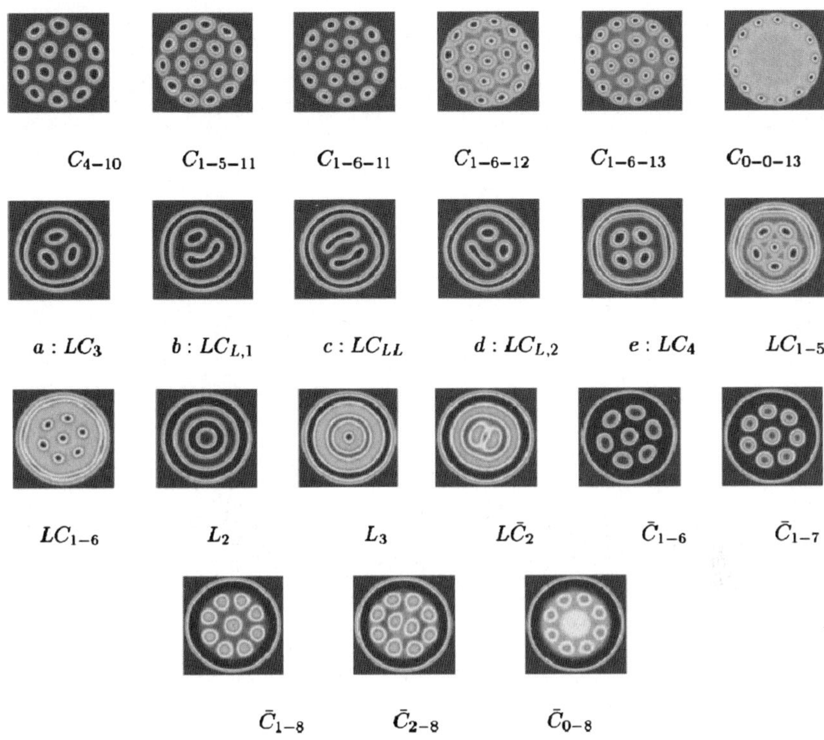

Fig. 25 Monomer density plots of the 21 nanostructured phases formed in a cylindrical pore with a D_p-value of 17 times the radius of gyration. The color ranges from *deep red* (A-rich regions) to *deep blue* (B-rich regions). The region outside the cylindrical pore is also colored deep blue. The notation C is used for cylindrical phases, L for lamellar phases, and LC for structures containing both lamellae and cylinders (intermediate structure). When the cylinders are composed of minority B component, an overbar is used. The number of cylinders in each ring, out from the center of the pore, is indicated by subscripts C_{i-j-k}. The number of L subscripts in the notation $LC_{L...L,i-j}$ indicates the number of lamellar segments in the inner region of a given intermediate structure. The second subscript indicates the number of cylinders of the minority species in the pore and whether these cylinders are arranged in rings. The stability regions for these structures are labeled on the phase diagram in Fig. 26. The notation (a, b, c, d, e) is used to label some of the intermediate phases in Fig. 26. Reproduced from [202]. © (2006) American Chemical Society

diagram for a diblock copolymer melt confined to a cylindrical pore with a D_p-value set to 17 times the radius of gyration of the model BCP in the $f - \chi N$ plane, where f is the volume fraction of one of the components, N the degree of polymerization of the BCP and χ the Flory–Huggins interaction parameter for the components of the BCP [202]. Figure 25 shows snapshots of the morphology types, and Fig. 26 the areas they occupy in the $f - \chi N$ plane. As compared to bulk systems and even thin film configura-

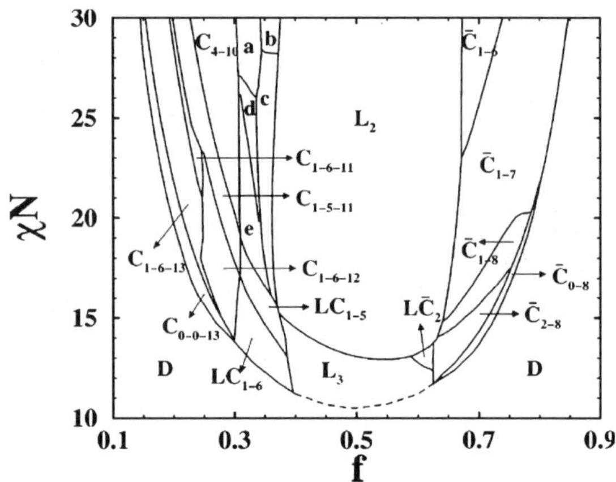

Fig. 26 Phase diagram for a diblock copolymer melt confined in a cylindrical pore with a D_p-value of 17 times the radius of gyration. The degree of polymerization of the copolymer is N, the Flory–Huggins parameter is χ, and f is the A monomer fraction. The disordered phase is labeled D; the other labels are as in Fig. 25. The *dashed curve* is an interpolation of the order–disorder transition curve. Reproduced from [202]. © (2006) American Chemical Society

tions, the phase behavior of BCPs under cylindrical confinement is much more complex.

Whereas the number of experimental publications dealing with sol-gel chemistry inside porous hard templates using BCP soft templates and that of theoretical papers dealing with self-assembly of BCP melts under cylindrical confinement is already large and increasing further, only a limited number of experimental reports on the self-assembly of BCP melts under cylindrical confinement are available up to now. However, BCP melts are an ideal system to explore the accessible morphology space since no fixation of non-equilibrium structures by gelation occurs, as in the case of sol-gel methods, and equilibrium structures can be obtained by thermal annealing or solvent annealing. Topographically patterned substrates have been used to increase the degree of order in thin BCP films confined to grooves or mesas [219–221]. However, up to now, only very few experimental studies of the structure evolution of BCPs under a true cylindrical confinement have been published. Russell and coworkers investigated PS-b-PBD diblock copolymers as a model system with highly repulsive blocks [74, 222]. In the case of symmetric PS-b-PBD that forms lamella in the bulk, concentric lamellae along the axes of the hard template pores were found. The PBD preferentially segregated to the pore walls, and alternating PS and PBD layers occurred and were visualized by TEM on ultrathin sections of the nanofibers in which the PBD was selectively stained (Fig. 27A). For ratios D_p/L_0 larger than 3.2 an outermost PDB layer was generally observed. A decrease in D_p led to a smaller number of concentric layers, whereas in the center either PS or PBD was located, de-

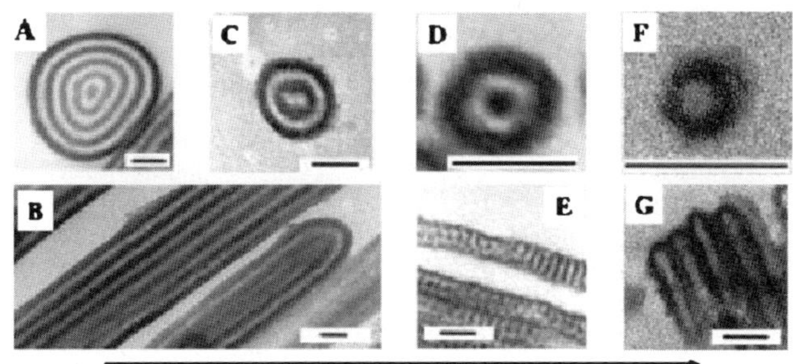

Fig. 27 TEM cross-sectional images of bulk lamella-forming PS-*b*-PBD confined in cylindrical pores. **A, C, D, F** Views across pore; **B, E, G** Views along pore; **A, B** $D_p/L_0 > 3.2$; **C** $D_p/L_0 = 3.2$; **D, E** $D_p/L_0 = 2.6$; and **F, G** $D_p/L_0 > 1.9$. *Scale bars*, 50 nm. Reproduced from [222]. © (2005) Wiley-VCH

pending on D_p (Fig. 27B,C). A deviation of the apparent repeat period from L_0 was found that increased for a given D_p-value towards the center of the pores where the layers exhibited the largest curvature, and for decreasing D_p-values. If D_p is comparable or even smaller than L_0 and D_p/L_0 is not an integer, the period of the BCP and the pore diameter are incommensurate. For $D_p/L_0 \approx 2.6$ a transition from a lamellar to a stacked disc or torus-type structure occurred. Normal to the rod axis, concentric layers were observed with PBD located at the centers and walls of the nanofibers. Along the axes of the nanorods, a stacked PS lamellar structure with a central spine and outer edges of PBD occurred (Fig. 27D,E) [223]. This morphology type has no counterparts in bulk systems but is in line with some of the theoretical predictions reviewed above. Further decreasing D_p/L_0 to a value of ≈ 1.9 resulted in another morphology transition. The pores of the hard template accommodated only one period, and the nanofibers consisted of a central core of PS, surrounded by a layer of PBD (Fig. 27F,G). This morphology requires a significant deformation of the BCP chains. However, concentric layers along the pores were apparently stabilized by the strong immiscibility of PS and PBD and favorable interfacial interactions of PBD with the pore walls [223].

In the case of asymmetric PS-*b*-PBD forming cylinders in the bulk, circular PBD layers parallel to the long axes of the hard template pores surround an area in which hexagonally packed PBD cylinders oriented along the pore axes are embedded in a PS matrix (Fig. 28A,B). However, the shape and size of the pores place constraints on the packing and both symmetry and separation distance of the domains were altered [74]. A decrease in D_p resulted in fewer cylinders within the cross-sectional area of the BCP nanofibers (Fig. 28C). Further decrease in D_p to values of 56–66 nm ($D_p/L_0 \approx 1.9$–2.3) led to the formation of a single cylindrical PBD domain in the center of the nanofibers with a PBD rim contacting the pore wall (Fig. 28D). The interfaces between the PBD center and rim and the PS showed undulations arising from the severe

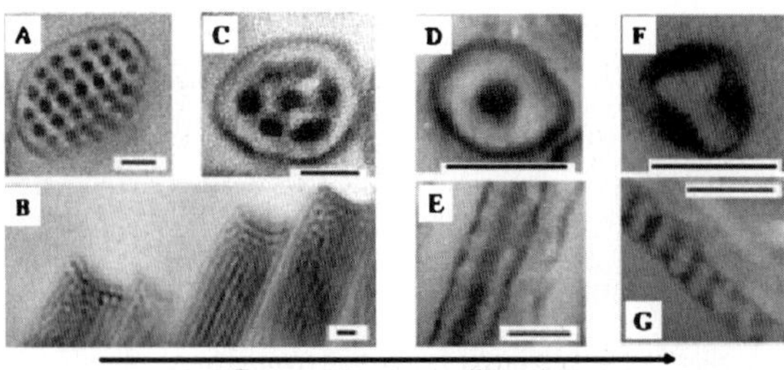

Fig. 28 TEM cross-sectional images of bulk cylinder-forming PS-*b*-PBD confined in cylindrical pores. **A, C, D, F** Views across pore; **B, E, G** views along pore; **A, B** $D_p/L_0 > 4$; **C** $D_p/L_0 = 4$; **D, E** $D_p/L_0 = 1.9$–2.3; and **F, G** $D_p/L_0 = 1.1$–1.5. *Scale bars*, 50 nm. Reproduced from [222]. © (2005) Wiley-VCH

geometric confinement, which were interpreted as a precursor to a change in the morphology (Fig. 28E) [222]. For D_p-values in the range from 33 to 45 nm ($D_p/L_0 \approx 1.1$–1.5), the PBD domain was still located at the pore walls, highlighting the edges of the nanofibers. However, instead of cylindrical domains aligned along the hard template pore axis, helical PBD structures maintaining contact with the pore walls were found whose pitch of 30 nm was close to L_0 (Fig. 28F,G) [224].

In another study, the diameter dependence of the morphology of symmetric PS-*b*-PMMA forming lamellae in the bulk ($L_0 = 42$ nm) on D_p was investigated [75]. In contrast to PS-*b*-PBD, PS-*b*-PMMA is only weakly segregated. However, the morphologies identified were similar to those obtained in the case of symmetric PS-*b*-PBD. Using AAO hard templates with D_p-values of 400, 180, 60 and 25 nm, a successive decrease in the number of concentric layers oriented parallel to the long axes of the hard template pores was found. For a D_p-value of 60 nm, a single circular PS layer surrounded a PMMA cylinder in the center of the nanofibers, and for a D_p-value of 25 nm, apparently a PS core was surrounded by a PMMA shell.

The morphologies of diblock copolymers under cylindrical confinement reported in experimental works are in line with the theoretical predictions reviewed above. However, because of the small number of publications addressing this issue, only a limited range of the anticipated morphology space has been explored. For example, no systematic variation of the surface field has been realized by modifying the pore walls of the hard template in such a way that their character changes form non-neutral to neutral. The limited interest in the investigation of the self-assembly of BCP melts inside hard templates might be due to the fact that, in contrast to the fabrication of mesoporous nanofibers by means of sol-gel chemistry, no obvious application exists. However, BCP nanofibers with complex morphologies consist of chemically distinct blocks. Therefore, it should be possible to realize selective segregation of precursors into one of the domains, a strategy that has successfully been applied in the case of thin film configurations. Therefore, BCP nanofibers having a complex morphology should have great potential as a soft template system in the synthesis of advanced functional one-dimensional nanostructures.

6
Multilayer Nanotubes by Layer-by-Layer Deposition

In the previous sections, approaches to the generation of mesoscopic fine structures in tubular but also solid one-dimensional nanostructures were discussed that involved deposition of target materials or precursors thereof into porous hard templates by a single infiltration step. Subsequently, the supramolecular organization is guided by the geometric confinement and interfacial interactions with the pore walls. Layer-by-layer deposition [225–227] is a generic access to nanoscopic multilayer systems with controlled composition that differs from single-step infiltration methods in that a series of successive deposition steps is performed, the number of which determines the properties of the nanotubes thus obtained to a large extent. To attach a new layer to the layers already deposited, specific interactions between the involved species or molecular recognition mechanisms are exploited. Initially, the layer-by-layer technique was applied to consecutively deposit oppositely charged polymeric polyelectrolytes from diluted solutions. Whereas the electrostatic repulsion between equally charged species limits the thickness of a deposited layer, the electrostatic attraction between the alternating oppositely charged layers is the glue holding together the entire assembly. The number of deposition steps determines the number of bilayers formed and therefore the thickness of the entire multilayer. Moreover, it is possible to incorporate inorganic nanoparticles if their surfaces are charged [228, 229]. Therefore, layer-by-layer assembly allows fabricating nanoscopic functional multilayer systems with outstandingly high precision. Whereas at first smooth substrates had been functionalized in this way, the coating of colloidal polymer particles with a multilayer structure consisting of silica nanoparticles and polymeric polyelectrolytes was reported by Caruso et al. in 1998 [230]. It was further shown that hollow capsules with walls consisting of polymeric multilayers can be prepared if polymeric colloidal particles are used as sacrificial templates [231]. Calcination of colloidal polymer particles covered by a multilayer structure in which silica nanoparticles were incorporated led to the formation of hollow silica capsules [232]. Several excellent reviews deal with the fabrication of free-standing and three-dimensional nanostructures by layer-by-layer deposition [233–235].

It appears to be straightforward to fabricate polymeric tubular structures with complex but well-controlled wall morphologies and adjustable wall thickness by performing layer-by-layer deposition into porous hard templates. To this end, Ai et al. employed a pressure-filter-template technique to deposit poly(allylamine hydrochloride as the anionic and poly(styrenesulfonic acid) as the cationic component from aqueous solutions also containing NaCl into AAO with a D_p-value of about 200 nm. After starting the deposition sequence with the formation of a poly(allylamine hydrochloride) layer directly on the pore walls, stable but flexible nanotubes consisting of three bilayers with a wall thickness of 50–80 nm were obtained even after etching the hard template with aqueous NaOH solution [236]. The thickness of the nanotube walls was one order of magnitude larger than that of corresponding multilayer structures prepared on smooth substrates in which a bilayer has a thickness of a few nanometers. Using PC membranes with a D_p-value of 400 nm and a T_p-value of 10 μm whose walls were initially coated with poly(ethylenimine), poly(acrylic acid)/poly(allylamine hydrochloride) multilayers were deposited onto the pore walls in the presence of Cu^{2+} and then thermally cross-linked. Moreover, positively charged Au nanoparticles were incorporated in nanotube walls in alternation with four-layer polyelectrolyte structures, and negatively charged semiconductor nanoparticles in alternation with three-layer polyelectrolyte structures. Whereas the wall thickness of the nanotubes thus obtained, which was of the order of several tens of nanometers, could be adjusted by the number of successive deposition cycles, the functionality of the embedded inorganic nanoparticles was preserved [237]. The wall thicknesses of the nanotubes reported in this study were only slightly lager than those in smooth configurations, and the mechanical stability of the nanotubes depended on the number of bilayers their walls consisted of.

Layer-by-layer deposition into porous hard templates has meanwhile been extended to other polyelectrolyte pairs. For example, Ai et al. prepared polypyrrole/poly(allylamine hydrochloride) nanotubes consisting of six or 12 bilayers in PC membranes with a D_p-value of 400 nm that had initially been coated with poly(ethylenimine) [238]. Again, the observed value of the wall thickness of the nanotubes of some tens of nanometers was much larger than that of corresponding multilayer systems deposited on smooth substrates. However, a clear dependence of the wall thickness on the number of deposition cycles was found. Nanotubes consisting of dendrimers were fabricated by Kim and coworkers (Fig. 29a) [239]. Dendrimers, synthesized by stepwisely attaching another generation of low-molar mass building blocks to a parent structure, represent a class of functional materials which can be customized with an unrivaled precision. On the one hand, they can be employed as functional nanocontainers. On the other hand, they contain a well-defined number of terminal functional groups residing at their surface [240–242]. For the preparation of the dendrimer nanotubes, bilayers containing globular-shaped, N,N-disubstituted hydrazine phosphorus-containing dendrimers [243] of the fourth generation having 96 terminal functional groups with either cationic $[G_4(NH^+Et_2Cl^-)_{96}]$ or anionic $[G_4(CH-COO^-Na^+)_{96}]$

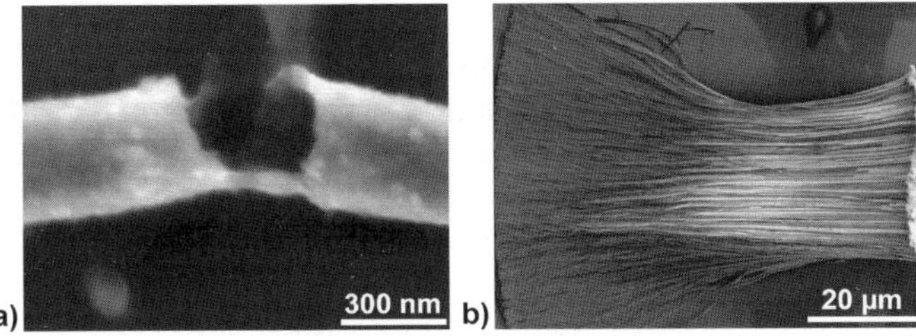

Fig. 29 SEM images of dendrimer nanotubes obtained by layer-by-layer deposition. **a** Broken nanotube; **b** array consisting of nanotubes exhibiting a gradient of their mechanical stability along their long axes. Reproduced from [239]. © (2005) Wiley-VCH

character were deposited on the walls of AAO hard templates with a D_p-value of 400 nm. Since dendrimers can be considered as hard spheres, nanotubes consisting of dendrimeric polyelectrolytes might be useful if swelling or deswelling needs to be minimized. The mechanical stability of the dendrimer nanotubes increased with the number of deposited bilayers. In the case of dendrimer nanotubes with high aspect ratios of the order of 200 that were prepared in hard templates with closed pore bottoms, their mechanical stability decreased with increasing distance to the pore opening. Whereas the nanotube segments initially located next to the pore openings with a length of about 35 μm were rigid, the nanotube segments farther away from the pore openings were prone to mechanical deformation due to the occurrence of capillary forces, which occur when nanofiber arrays dry after the wet-chemical etching of the hard template (Fig. 29b). Lu et al. reported the deposition of the hemoprotein human serum albumin (HSA) into AAO hard templates with a D_p-value of about 200 nm [244, 245]. Inversion of the charges borne by the HSA molecules was achieved by adjusting the pH-value of the solutions used for deposition to values below or above the isoelectric point of HSA. Therefore, in alternating deposition steps HSA could be deposited as a polycation or as a polyanion. Moreover, nanotubes whose walls consisted of HSA/phosholipid multilayers were prepared [244].

Extending the initial approach to exploit electrostatic interactions between polyelectrolytic building blocks for their rational arrangement in nanotube walls by layer-by-layer assembly, a number of modifications of this methodology based on different kinds of interactions have been developed. For example, the formation of hydrogen bonds between hydroxyl groups of poly(acrylic acid) and the nitrogen groups of poly(4-vinylpyridine) was exploited to coat the pore walls of PC membranes with multilayer structures consisting of these polyelectrolytes [246]. An advantage of this approach lies in the fact that solutions in organic solvents can be used to deposit the monolayers. Nanotubes consisting of poly(ethylenimine)/poly(styrene-*alt*-maleic anhydride) multilayers were obtained by connecting the alternating monolayers by amide bonds [247]. A similar approach was applied to incorporate the fluorescent compound 3,4,9,10-perylenetetracarboxylicdianhydride into nanotubes having multilayered walls with poly(ethylenimine) as a second component [248]. Since the fluorescent component retained its fluorescence, single nanotubes could be imaged by fluorescence microscopy, and the formation of the multilayer structure could be monitored by UV absorption spectroscopy. It was found that the absorption linearly increased along with the number of deposition cycles.

Hou et al. reported the fabrication of glutaraldehyde/protein nanotubes [249]. Using phosphorous-containing coupling agents, a first glutaraldehyde layer was grafted onto the pore walls of AAO hard templates with a D_p-value of 200 nm. Subsequently, a protein layer was bonded with its free amino sites to the excess aldehyde functions of the glutaraldehyde layer, and in turn another glutaraldehyde layer to free amino sites of the proteins. For example, bioactive nanotubes could be fabricated by the repeated deposition of glutaraldehyde/glucose oxidase bilayers. The activity of the glucose oxidase in the liberated nanotubes increased along with the number of protein layers in their walls. However, inside the AAO hard templates the activity of the nanotubes decreased for more than three bilayers, because the accessibility of the protein molecules through the hollow channel inside the nanotubes became more and more limited as the diameter of the channel decreased with each additional layer. Also, Hou et al. showed that hemoglobin nanotubes that were produced in a similar manner exhibited heme electroactivity. Tian et al. used AAO hard templates activated with a poly(ethylenimine)/poly(sodium-4-styrenesulfonate) bilayer to deposit cytochrome C/glutaraldehyde bilayers and obtained nanotubes in which the bioactivity and the electronic properties of cytochrome C were retained [250]. Hou et al. fabricated DNA nanotubes employing a hybridization-based layer-by-layer strategy. After the initial grafting of DNA strands on the pore walls with the aid of phosphorous-containing coupling agents, the hard templates were successively immersed into DNA solutions, allowing binding of further DNA strands to those already immobilized by hybridization [251].

Whereas layer-by-layer deposition into porous hard templates has been proven to be a promising access to precisely designed polymeric nanotubes that can functionalize hard templates or that can be released, two problems still need to be addressed. The first one is elucidating the structure formation of the polyelectrolyte layers inside hard templates. The significantly increased thickness of bilayers reported by Ai et al. [236] was also observed by other authors. Lee et al. found that the thickness of poly(allylamine hydrochloride)/poly(sodium-4-styrenesulfonate) multilayers deposited into porous PC membranes exceeded that of corresponding multilayers on smooth substrates obtained after the same number of deposition cycles. For example, 24.5 bilayers had a thickness of 250 nm within a hard template as compared to 155 nm on a smooth silicon wafer [252]. Alem et al. studied the layer-by-layer deposition of a pair of strong polyelectrolytes, namely poly(vinylbenzylammonium chloride) as a polycation and poly(sodium-4-styrenesulfonate) as a polyanion, into track-etched PC membranes with D_p-values ranging from 50 to 850 nm. The end-to-end distances of the polyelectrolyte chains were systematically varied by varying the molecular weight and the ionic strength of the solutions used for deposition [253]. Whereas the bilayers deposited on smooth substrates had a thickness of 1–3 nm, the first bilayer deposited into porous hard templates covering the pore walls had a thickness of 50–120 nm. Further deposition cycles led to only small increases in the thickness of the polyelectrolyte layers, which turned out to be nearly independent of the end-to-end distance of the polyelectrolytes and the ionic strength of the stock solutions used but strongly depended on the D_p-values of the hard templates. For small D_p-values, the thickness of the polyelectrolyte layers was proportional to D_p, whereas progressive deviations from this relationship were found for D_p-values larger than 250 nm. On the basis of geometric considerations, Alem et al. proposed a mechanism governing the growth of polyelectrolyte layers inside porous hard templates that involves the enrichment of the polyelectrolytes inside the pores. Hence, a dense, swollen polyelectrolyte gel fills pores and collapses upon drying.

The second issue that needs to be further addressed is the development of strategies for the anchoring of the first deposited layer onto the pore walls of the hard templates. This is particularly the case for the widely used AAO membranes, whose pore walls consist of amorphous alumina containing water, electrolyte anions and positively charged defects (Sect. 2). Moreover, composition and distribution of the contaminations across the pore walls are inhomogeneous (see, for example, [30–32]). Therefore, isotropic etching steps performed to widen the pores of as-anodized, self-ordered AAO with an initial porosity of 10% [40] or below will change the properties of the pore walls and affect their reactivity. In the case of commercially available disordered AAO membranes with a D_p-value of 200 nm positively charged polyelectrolytes such as poly(ethylenimine) [247, 250] or human serum albumin at a pH value of 3.8 [244] could directly be deposited as the first layer. However, Dai et al. reported a procedure to coat the same type of hard templates that started with the deposition of poly(acrylic acid), hence with a polyanion, at a pH value of 4.0 [254]. The reports dealing with the surface chemistry of AAO are to a large extent inconsistent, and it appears that the surface properties of the pore walls largely depend on the anodization conditions and post-anodization treatments. Strategies to overcome the problems associated with the lack of knowledge of the properties of the hard templates are based on their modification by grafting anchor layers onto the pore walls. For example, Kim et al. used 3-aminopropyl-dimethylethoxysilane, a silane coupling agent, to generate a layer with a high density of positive charges on the walls of self-ordered porous alumina with a D_p-value of 400 nm [239]. Hou et al. [249, 251, 255] adapted a surface modification strategy based on a double layer of phosphorous-containing coupling agents initially introduced by Mallouk and coworkers [256] that allowed further layer-by-layer deposition mediated by specific chemical interactions.

First applications of polymeric nanotubes fabricated by layer-by-layer deposition have already been reported. Using track-etched PC membranes with D_p-values ranging from 400 to 800 nm functionalized with poly(allylamine hydrochloride)/poly(sodium-4-styrenesulfonate) bilayers, Lee et al. demonstrated reversible pH-induced hysteretic gating [252]. Membrane pores, the walls of which were covered with 18.5 bilayers, could be closed to a pH-value of 2.5 by swelling the polyelectrolytes. The pores thus closed retained their closed state up to pH 9. At higher pH values, the swollen polyelectrolyte layer collapsed, and the pores switched to the open state. The switching behavior of the system could be customized by the number of bilayers deposited on the pore walls. The flux of pH-adjusted water through membrane was studied and indicated discontinuous swelling/deswelling behavior (Fig. 30) but lesser swelling in pores than on smooth substrates (Fig. 31). Lee et al. suggested that swelling in the pores of hard templates is suppressed because of the curvature-induced stress generated by the volume expansion in a curved geometry.

Dai et al. reported a strategy for analyzing proteins by selective binding to antibodies in such a way that nonspecific adsorption and protein denaturation could be prevented. To this end, poly(acrylic acid)/protonated poly(allylamine) multilayers that are well known to resist nonspecific adsorption and to allow for covalent immobilization of arrays of active antibodies were coated on the walls of AAO hard templates with

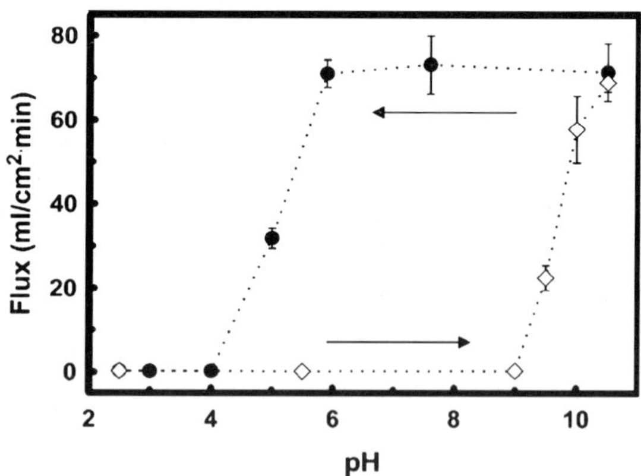

Fig. 30 Reversible pH-induced hysteretic gating with layer-by-layer nanotubes. Changes in flux through a membrane functionalized with poly(allylamine hydrochloride)/poly(sodium-4-styrenesulfonate) nanotubes as a function of pH. The *filled circles* and *open diamonds* represent data generated after a pH 10.5 pretreatment and after a pH 2.5 pretreatment, respectively. *Error bars* represent standard deviations. Reproduced from [252]. © (2006) American Chemical Society

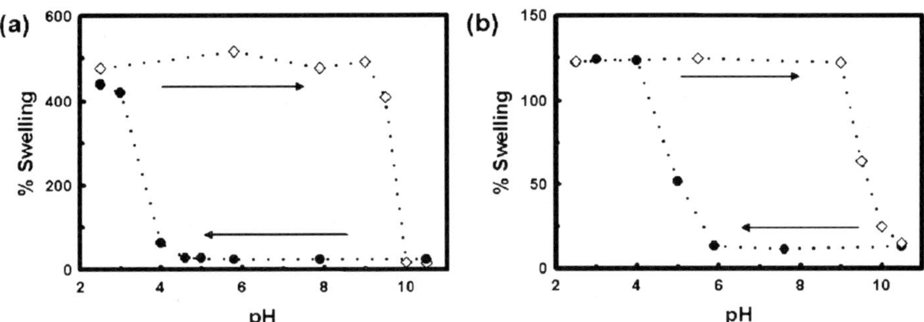

Fig. 31 Comparison of the swelling ratio of poly(allylamine hydrochloride)/poly(sodium-4-styrenesulfonate) multilayers **a** on a planar Si substrate (determined by in-situ ellipsometry) and **b** in the cylindrical pores of a 800 nm pore track-etched PC membrane. The *filled circles* and *open diamonds* represent data generated after a pH 10.5 pretreatment and after a pH 2.5 pretreatment, respectively. Reproduced from [252]. © (2006) American Chemical Society

a D_p-value of 200 nm. Activation of the surface carboxyl groups of the poly(acrylic acid) with N-(3-dimethylaminopropyl)-N'-ethylcarbodiimide and N-hydroxysuccinimide enabled the covalent attachment of antibodies. A 500-fold increase in the surface area as compared to thin film configurations decreased the protein-microarray detection limit by two orders of magnitude [254]. Feng et al. adapted the procedure for the fabrication of dendrimer nanotubes reported by Kim et al. [239] to incorporate a graded bandgap structure similar to that previously reported by Franzl et al. for thin-film configurations on smooth substrates [257] into the walls of dendrimer nanotubes. The dendrimer layers acted as a rigid scaffold for the engineering of a multilayer configuration of inorganic semiconductor quantum dots having different diameters. Taking advantage of a fluorescence resonance energy transfer cascade from donor nanoparticles located near the outer surface on the nanotube walls to acceptor particles located near the inner surface of the nanotube walls, the hybridization of DNA strands grafted on the inner tube surface with complementary labeled DNA strands could be detected with significantly increased sensitivity [258].

7
Conclusion

One-dimensional polymeric nanostructures or one-dimensional nanostructures derived from polymeric soft templates are being considered as functional building blocks for a broad range of device architectures, and some promising applications have already been reported. Various well-established approaches allow forming virtually any functional material into a tubular or a solid rod-like shape. This review is based on the paradigm that the control over their internal mesoscopic fine structure is the most crucial means of tailoring the properties of one-dimensional nanoobjects, whereas the presence or absence of a central hollow space and therefore the distinction between "tubes" and "solid rods" is, in many cases, of minor importance. This is all the more the case for polymeric nanotubes and nanorods as the supramolecular organization on mesoscopic length scales determines the properties of polymeric materials to a large extent. Nanoporous, shape-defining hard templates provide a two-dimensionally confined space in which self-organization processes such as crystallization, mesophase formation, and phase separation may result in supramolecular fine structures fundamentally different from those obtained in thin film configurations and in the bulk. A particular advantage of hard templates is the possibility to induce and manipulate self-assembly inside the shape-defining pores. Therefore, many more parameters allow tailoring of the mesoscopic morphology of the nanofibers in hard-template-based preparation processes than in procedures for their production not relying on the rigidity of confining pore walls. The supramolecular organization in the two-dimensional confinement of nanopores can be manipulated by the pore diameter, the nature of the pore walls, the composition of the infiltrated material, environmental conditions, and the thermal history of the sample. Moreover, nanotubes characterized by complex, multilayered walls are accessible by successive deposition steps into the hard templates, thus exploiting specific interactions between the deposited species and the material already deposited. Thus, by the preparative approaches reviewed in this contribution complex functional supramolecular structures can be generated in one-dimensional nanofibers. Examples for this are ordered mesoporous structures and microphase morphologies, uniform crystalline and liquid crystalline textures, as well as the incorporation of inorganic nanoparticles into the walls of polymer nanotubes. The control over the supramolecular organization of the materials the nanotubes consist of is the prerequisite for the rational tailoring of peculiar mechanical, optical and electronic properties.

Even though first applications for nanotubes with walls exhibiting a customized supramolecular organization have been demonstrated, the understanding of and the control over the relevant structure formation processes involving polymeric materials confined to hard templates need to be improved. Promising configurations for device components are, on the one hand, composite membranes consisting of the hard template and the nanotubes inside their pores, which are of potential interest for applications in the fields of storage, separation and catalysis. On the other hand, arrays of released nanofibers may have specific adhesive and wetting properties or can be used as nanosensor and nanoactuator arrays. However, the up-scaling of the production of such nanofiber assemblies requires their destruction-free removal from recyclable hard templates. To create nanofiber arrays usable in real life applications, their mechanical stability has to be improved and condensation of the nanofibers has to be prevented by optimizing the geometry of the array.

Acknowledgements Financial support by the German Research Foundation (SPP 1165 "Nanowires and Nanotubes"; DFG-NSF Materials World Network) is gratefully acknowledged.

References

1. Bognitzki M, Hou HQ, Ishaque M, Frese T, Hellwig M, Schwarte C, Schaper A, Wendorff JH, Greiner A (2000) Adv Mater 12:637
2. Wehrspohn RB (2005) Ordered porous nanostructures and applications. Springer, Berlin Heidelberg New York
3. Martin CR (1994) Science 266:1961
4. Martin CR (1996) Chem Mater 8:1739
5. Huczko A (2000) Appl Phys A 70:365
6. Cai ZH, Martin CR (1989) J Am Chem Soc 111:4138
7. Liang WB, Martin CR (1990) J Am Chem Soc 112:9666
8. Martin CR, Vandyke LS, Cai ZH, Liang WB (1990) J Am Chem Soc 112:8976
9. Martin CR (1991) Adv Mater 3:457
10. Parthasarathy RV, Martin CR (1994) Chem Mater 6:1627
11. Martin CR (1995) Acc Chem Res 28:61
12. Menon VP, Lei JT, Martin CR (1996) Chem Mater 8:2382
13. Cepak VM, Martin CR (1999) Chem Mater 11:1363
14. Penner RM, Martin CR (1986) J Electrochem Soc 133:2206
15. Kohli P, Harrell CC, Cao ZH, Gasparac R, Tan WH, Martin CR (2004) Science 305:984
16. Lakshmi BB, Martin CR (1997) Nature 388:758
17. Jin MH, Feng XJ, Feng L, Sun TL, Zhai J, Li TJ, Jiang L (2005) Adv Mater 17:1977
18. Xu JH, Li M, Zhao Y, Lu QH (2007) Coll Surf A 302:136
19. Chen X, Steinhart M, Hess C, Gösele U (2006) Adv Mater 18:2153
20. Grimm S, Schwirn K, Göring P, Knoll H, Miclea PT, Greiner A, Wendorff JH, Wehrspohn RB, Gösele U, Steinhart M (2007) Small 3:993
21. Ott AW, Klaus JW, Johnson JM, George SM, McCarley KC, Way JD (1997) Chem Mater 9:707
22. Xiong G, Elam JW, Feng H, Han CY, Wang HH, Iton LE, Curtiss LA, Pellin MJ, Kung M, Kung H, Stair PC (2005) J Phys Chem B 109:14059
23. Apel P (2001) Radiation Measurements 34:559
24. Crawford GP, Steele LM, Ondriscrawford R, Iannacchione GS, Yeager CJ, Doane JW, Finotello D (1992) J Chem Phys 96:7788
25. Diggle JW, Downie TC, Goulding CW (1969) Chem Rev 69:365
26. O'Sullivan JP, Wood GC (1970) Proc Royal Soc London Ser A 317:511
27. Keller F, Hunter MS, Robinson DL (1953) J Electrochem Soc 100:411
28. Thompson GE, Wood GC (1981) Nature 290:230
29. Thompson GE, Furneaux RC, Wood GC, Richardson JA, Goode JS (1978) Nature 272:433
30. Chen W, Yuan JH, Xia XH (2005) Anal Chem 77:8102
31. Choi J, Luo Y, Wehrspohn RB, Hillebrand R, Schilling J, Gösele U (2003) J Appl Phys 94:4757

32. Vrublevsky I, Jagminas A, Schreckenbach J, Goedel WA (2007) Appl Surf Sci 253:4680
33. Zhao L, Yosef M, Pippel E, Hofmeister H, Steinhart M, Gösele U, Schlecht S (2006) Angew Chem Int Ed 45:8042
34. Zhao LL, Yosef M, Steinhart M, Göring P, Hofmeister H, Gösele U, Schlecht S (2006) Angew Chem Int Ed 45:311
35. Wang Y, Wu K (2005) J Am Chem Soc 127:9686
36. Masuda H, Fukuda K (1995) Science 268:1466
37. Masuda H, Hasegwa F, Ono S (1997) J Electrochem Soc 144:L127
38. Masuda H, Yada K, Osaka A (1998) Japan J Appl Phys Part 2 37:L1340
39. Li AP, Müller F, Birner A, Nielsch K, Gösele U (1998) J Appl Phys 84:6023
40. Nielsch K, Choi J, Schwirn K, Wehrspohn RB, Gösele U (2002) Nano Lett 2:677
41. Lee W, Ji R, Gösele U, Nielsch K (2006) Nat Mater 5:741
42. Chu SZ, Wada K, Inoue S, Isogai M, Yasumori A (2005) Adv Mater 17:2115
43. Masuda H, Takenaka K, Ishii T, Nishio K (2006) Jpn J Appl Phys 2-Letters & Express Lett 45: L1165
44. Masuda H, Nagae M, Morikawa T, Nishio K (2006) Jpn J Appl Phys 2-Letters & Express Lett 45: L406
45. Masuda H, Yamada H, Satoh M, Asoh H, Nakao M, Tamamura T, Ye (1997) Appl Phys Lett 71:2770
46. Choi J, Nielsch K, Reiche M, Wehrspohn RB, Gösele U (2003) J Vacuum Sci Technol B 21:763
47. Lee W, Ji R, Ross CA, Gösele U, Nielsch K (2006) Small 2:978
48. Berg JC (1993) Wettability. Dekker, New York
49. Fox HW, Hare EF, Zisman WA (1955) J Phys Chem 59:1097
50. De Gennes PG (1985) Rev Modern Phys 57:827
51. Ausserré D, Picard AM, Léger L (1986) Phys Rev Lett 57:2671
52. Léger L, Joanny JF (1992) Reports Progr Phys 55:431
53. Derjaguin BV, Churaev NV (1974) J Coll Inter Sci 49:249
54. Mate CM, Novotny VJ (1991) J Chem Phys 94:8420
55. Israelachvili JN (1991) Intermolecular and surface forces. Academic Press, New York
56. Bernadiner MG (1998) Transport in Porous Media 30:251
57. Everett DH, Haynes JM (1972) J Coll Inter Sci 38:125
58. Hammond PS (1983) J Fluid Mechanics 137:363
59. Lenormand R (1990) J Phys-Condens Matter 2:SA79
60. Steinhart M, Wendorff JH, Greiner A, Wehrspohn RB, Nielsch K, Schilling J, Choi J, Gösele U (2002) Science 296:1997
61. Zhang MF, Dobriyal P, Chen JT, Russell TP, Olmo J, Merry A (2006) Nano Lett 6:1075
62. Steinhart M, Wendorff JH, Wehrspohn RB (2003) Chem Phys Chem 4:1171
63. Steinhart M, Wehrspohn RB, Gösele U, Wendorff JH (2004) Angew Chem Int Ed 43:1334
64. Steinhart M, Göring P, Dernaika H, Prabhukaran M, Gösele U, Hempel E, Thurn-Albrecht T (2006) Phys Rev Lett 97:027801
65. Steinhart M, Murano S, Schaper AK, Ogawa T, Tsuji M, Gösele U, Weder C, Wendorff JH (2005) Adv Functional Mater 15:1656
66. Lau S, Zheng RK, Chan HLW, Choy CL (2006) Mater Lett 60:2357
67. Zheng RK, Yang Y, Wang Y, Wang J, Chan HLW, Choy CL, Jin CG, Li XG (2005) Chem Commun, p 1447
68. She XL, Song GJ, Li JJ, Han P, Yang SJ, Peng Z (2006) J Mater Res 21:1209
69. She XL, Song GJ, Li JJ, Han P, Yang SJ, Wang SL, Peng Z (2006) Polym J 38:639
70. Kim E, Xia YN, Whitesides GM (1995) Nature 376:581

71. Suh KY, Kim YS, Lee HH (2001) Adv Mater 13:1386
72. Moon SI, McCarthy TJ (2003) Macromolecules 36:4253
73. Kriha O, Zhao LL, Pippel E, Gösele U, Wehrspohn RB, Wendorff JH, Steinhart M, Greiner A (2007) Adv Functional Mater 17:1327
74. Xiang HQ, Shin K, Kim T, Moon SI, McCarthy TJ, Russell TP (2004) Macromolecules 37:5660
75. Sun YM, Steinhart M, Zschech D, Adhikari R, Michler GH, Gösele U (2005) Macromolecular Rapid Commun 26:369
76. Shin K, Obukhov S, Chen J-T, Huh J, Hwang Y, Mok S, Dobriyal P, Thiyagarajan P, Russell TP (2007) Nat Mater 6:961
77. Song GJ, She XL, Fu ZF, Li JJ (2004) J Mater Res 19:3324
78. Li JJ, Song GJ, She XL, Han P, Peng Z, Chen D (2006) Polymer J 38:554
79. Steinhart M, Senz S, Wehrspohn RB, Gösele U, Wendorff JH (2003) Macromolecules 36:3646
80. Primak SV, Jin T, Dagger AC, Finotello D, Mann EK (2002) Phys Rev E 65:031804
81. Ai SF, Cui Y, He Q, Tao C, Li JB (2006) Colloid Surface A 275:218
82. Binder K (1998) J Non-Equil Thermod 23:1
83. Gelb LD, Gubbins KE, Radhakrishnan R, Sliwinska-Bartkowiak M (1999) Rep Prog Phys 62:1573
84. Liu AJ, Durian DJ, Herbolzheimer E, Safran SA (1990) Phys Rev Lett 65:1897
85. Steinhart M, Jia ZH, Schaper AK, Wehrspohn RB, Gösele U, Wendorff JH (2003) Adv Mater 15:706
86. De Gennes PG (1981) Macromolecules 14:1637
87. Lai PY, Binder K (1992) J Chem Phys 97:586
88. O'Shaughnessy B, Vavylonis D (2005) J Phys-Condens Matter 17:R63
89. Chen JT, Shin K, Leiston-Belanger JM, Zhang MF, Russell TP (2006) Adv Functional Mater 16:1476
90. Rodriguez AT, Chen M, Chen Z, Brinker CJ, Fan HY (2006) J Am Chem Soc 128:9276
91. Gonuguntla M, Sharma A (2004) Langmuir 20:3456
92. Rayleigh L (1892) Philosophical Magazine Serié 5 34:177
93. Quere D, Dimeglio JM, Brochard-Wyart F (1990) Science 249:1256
94. Chen JT, Zhang MF, Russell TP (2007) Nano Letters 7:183
95. Sirringhaus H, Brown PJ, Friend RH, Nielsen MM, Bechgaard K, Langeveld-Voss BMW, Spiering AJH, Janssen RAJ, Meijer EW, Herwig P, de Leeuw DM (1999) Nature 401:685
96. Lovinger AJ (1983) Science 220:1115
97. Kepler RG, Anderson RA (1978) J Appl Phys 49:1232
98. Dveyaharon H, Sluckin TJ, Taylor PL, Hopfinger AJ (1980) Phys Rev B 21:3700
99. Broadhurst MG, Davis GT (1981) Ferroelectrics 32:177
100. Shin K, Woo E, Jeong YG, Kim C, Huh J, Kim KW (2007) Macromolecules 40:6617
101. Wu H, Wang W, Yang HX, Su ZH (2007) Macromolecules 40:4244
102. Zhi LJ, Gorelik T, Wu JS, Kolb U, Müllen K (2005) J Am Chem Soc 127:12792
103. Zhi LJ, Wu JS, Li JX, Kolb U, Müllen K (2005) Angew Chem Int Ed 44:2120
104. Woo E, Huh J, Jeong YG, Shin K (2007) Phys Rev Lett 98:136103
105. Koenig JL (1999) Spectroscopy of polymers. Elsevier, Amsterdam
106. Martin CR, Parthasarathy R, Menon V (1993) Synthetic Metals 55:1165
107. Cai ZH, Lei JT, Liang WB, Menon V, Martin CR (1991) Chem Mater 3:960
108. Keller A (1957) Philosophical Magazine 2:1171
109. Reiter G, Strobl GR (2007) Progress in understanding of polymer crystallization. Springer, Berlin Heidelberg New York

110. Strobl GR (2007) The physics of polymers: concepts for understanding their structures and behavior. Springer, Berlin Heidelberg New York
111. Keith HD, Padden FJ (1963) J Appl Phys 34:2409
112. Bassett DC, Vaughan AS (1985) Polymer 26:717
113. Granasy L, Pusztai T, Tegze G, Warren JA, Douglas JF (2005) Phys Rev E 72:011605
114. Turnbull D (1950) J Appl Phys 21:1022
115. Turnbull D (1950) Journal of Chemical Physics 18:198
116. O'Carroll D, Lieberwirth I, Redmond G (2007) Small 3:1178
117. Quiram DJ, Register RA, Marchand GR, Adamson DH (1998) Macromolecules 31:4891
118. Loo YL, Register RA, Ryan AJ (2000) Phys Rev Lett 84:4120
119. Zhu L, Cheng SZD, Calhoun BH, Ge Q, Quirk RP, Thomas EL, Hsiao BS, Yeh FJ, Lotz B (2000) J Am Chem Soc 122:5957
120. Loo YL, Register RA, Ryan AJ, Dee GT (2001) Macromolecules 34:8968
121. Reiter G, Castelein G, Sommer JU, Röttele A, Thurn-Albrecht T (2001) Phys Rev Lett 8722:226101
122. Beiner M, Rengarajan GT, Pankaj S, Enke D, Steinhart M (2007) Nano Lett 7:1381
123. Zhao LL, Lu TZ, Yosef M, Steinhart M, Zacharias M, Gösele U, Schlecht S (2006) Chem Mater 18:6094
124. Laschat S, Baro A, Steinke N, Giesselmann F, Hagele C, Scalia G, Judele R, Kapatsina E, Sauer S, Schreivogel A, Tosoni M (2007) Angew Chem Int Ed 46:4832
125. Pisula W, Tomovic Z, El Hamaoui B, Watson MD, Pakula T, Müllen K (2005) Adv Functional Mater 15:893
126. Wu JS, Pisula W, Müllen K (2007) Chem Rev 107:718
127. Pisula W, Kastler M, Wasserfallen D, Davies RJ, Garcia-Gutierrez MC, Müllen K (2006) J Am Chem Soc 128:14424
128. Kastler M, Pisula W, Davies RJ, Gorelik T, Kolb U, Müllen K (2007) Small 3:1438
129. Zhi LJ, Wu JS, Li JX, Stepputat M, Kolb U, Müllen K (2005) Adv Mater 17:1492
130. Liu QY, Li Y, Liu HG, Chen YL, Wang XY, Zhang YX, Li XY, Jiang JZ (2007) J Phys Chem C 111:7298
131. Barrett C, Iacopino D, O'Carroll D, De Marzi G, Tanner DA, Quinn AJ, Redmond G (2007) Chem Mater 19:338
132. Palermo V, Liscio A, Talarico AM, Zhi LJ, Mullen K, Samori P (2007) Philos T Roy Soc A 365:1577
133. Enke D, Janowski F, Schwieger W (2003) Micropor Mesopor Mater 60:19
134. Bognitzki M, Czado W, Frese T, Schaper A, Hellwig M, Steinhart M, Greiner A, Wendorff JH (2001) Adv Mater 13:70
135. Bognitzki M, Frese T, Steinhart M, Greiner A, Wendorff JH, Schaper A, Hellwig M (2001) Poly Eng Sci 41:982
136. Greiner A, Wendorff JH (2007) Angew Chem Int Ed 46:5670
137. Cahn JW (1961) Acta Metallurgica 9:795
138. Hashimoto T (1993) In: Thomas EL (ed) Structure and Properties of Polymers (Materials Science and Technology), vol 12. Wiley, Weinheim, p 251
139. De Gennes PG (1980) J Chem Phys 72:4756
140. Jones RAL, Norton LJ, Kramer EJ, Bates FS, Wiltzius P (1991) Phys Rev Lett 66:1326
141. Krausch G, Dai CA, Kramer EJ, Bates FS (1994) Berichte Der Bunsen-Gesellschaft-Physical Chemistry Chemical Physics 98:446
142. Sung L, Karim A, Douglas JF, Han CC (1996) Phys Rev Lett 76:4368
143. Albano EV, Binder K, Heermann DW, Paul W (1992) Physica A 183:130
144. Gelb LD, Gubbins KE (1997) Phys Rev E 55:R1290

145. Gelb LD, Gubbins KE (1997) Phys Rev E 56:3185
146. Gelb LD, Gubbins KE (1997) Physica A 244:112
147. Luo Y, Lee SK, Hofmeister H, Steinhart M, Gösele U (2004) Nano Lett 4:143
148. Nielsch K, Castano FJ, Ross CA, Krishnan R (2005) J Appl Phys 98:6
149. Nielsch K, Castano FJ, Matthias S, Lee W, Ross CA (2005) Adv Eng Mater 7:217
150. Tanaka K, Takahara A, Kajiyama T (1998) Macromolecules 31:863
151. Tanaka K, Kajiyama T, Takahara A, Tasaki S (2002) Macromolecules 35:4702
152. Walheim S, Böltau M, Mlynek J, Krausch G, Steiner U (1997) Macromolecules 30:4995
153. Buck E, Fuhrmann J (2001) Macromolecules 34:2172
154. Harris M, Appel G, Ade H (2003) Macromolecules 36:3307
155. Kresge CT, Leonowicz ME, Roth WJ, Vartuli JC, Beck JS (1992) Nature 359:710
156. Attard GS, Glyde JC, Göltner CG (1995) Nature 378:366
157. Lu YF, Ganguli R, Drewien CA, Anderson MT, Brinker CJ, Gong WL, Guo YX, Soyez H, Dunn B, Huang MH, Zink JI (1997) Nature 389:364
158. Templin M, Franck A, DuChesne A, Leist H, Zhang YM, Ulrich R, Schädler V, Wiesner U (1997) Science 278:1795
159. Zhao DY, Feng JL, Huo QS, Melosh N, Fredrickson GH, Chmelka BF, Stucky GD (1998) Science 279:548
160. Joo SH, Choi SJ, Oh I, Kwak J, Liu Z, Terasaki O, Ryoo R (2001) Nature 412:169
161. Ryoo R, Joo SH, Kruk M, Jaroniec M (2001) Adv Mater 13:677
162. Liang CD, Hong KL, Guiochon GA, Mays JW, Dai S (2004) Angew Chem Int Ed 43:5785
163. Zhang FQ, Meng Y, Gu D, Yan Y, Yu CZ, Tu B, Zhao DY (2005) J Am Chem Soc 127:13508
164. Liang CD, Dai S (2006) J Am Chem Soc 128:5316
165. Zhang FQ, Gu D, Yu T, Zhang F, Xie SH, Zhang LJ, Deng YH, Wan Y, Tu B, Zhao DY (2007) J Am Chem Soc 129:7746
166. Ying JY, Mehnert CP, Wong MS (1999) Angew Chem Int Ed 38:56
167. Davis ME (2002) Nature 417:813
168. Kickelbick G (2005) Small 1:168
169. Lu AH, Schüth F (2006) Adv Mater 18:1793
170. Kanatzidis MG (2007) Adv Mater 19:1165
171. Wan Y, Shi YF, Zhao DY (2007) Chem Comm: 897
172. Wan Y, Zhao DY (2007) Chem Rev 107:2821
173. Yang ZL, Niu ZW, Cao XY, Yang ZZ, Lu YF, Hu ZB, Han CC (2003) Angew Chem Int Ed 42:4201
174. Liang ZJ, Susha AS (2004) Chem-Eur J 10:4910
175. Yao B, Fleming D, Morris MA, Lawrence SE (2004) Chem Mater 16:4851
176. Zhu WP, Han YC, An LJ (2005) Micropor Mesopor Mater 84:69
177. Chae WS, Lee SW, An MJ, Choi KH, Moon SW, Zin WC, Jung JS, Kim YR (2005) Chem Mater 17:5651
178. Lu QY, Gao F, Komarneni S, Mallouk TE (2004) J Am Chem Soc 126:8650
179. Wang DH, Kou R, Yang ZL, He JB, Yang ZZ, Lu YF (2005) Chem Commun, p 166
180. Platschek B, Petkov N, Bein T (2006) Angew Chem Int Ed 45:1134
181. Jin KW, Yao BD, Wang N (2005) Chem Phys Lett 409:172
182. Wu YY, Cheng GS, Katsov K, Sides SW, Wang JF, Tang J, Fredrickson GH, Moskovits M, Stucky GD (2004) Nat Mater 3:816
183. Wu YY, Livneh T, Zhang YX, Cheng GS, Wang JF, Tang J, Moskovits M, Stucky GD (2004) Nano Lett 4:2337

184. Yu K, Hurd AJ, Eisenberg A, Brinker CJ (2001) Langmuir 17:7961
185. Chen X, Knez M, Berger A, Nielsch K, Gösele U, Steinhart M (2007) Angew Chem Int Ed 46:6829
186. Chae WS, Lee SW, Kim YR (2005) Chem Mater 17:3072
187. Wang K, Wei MD, Morris MA, Zhou HS, Holmes JD (2007) Adv Mater 19:3016
188. Cott DJ, Petkov N, Morris MA, Platschek B, Bein T, Holmes JD (2006) J Am Chem Soc 128:3920
189. Steinhart M, Liang CD, Lynn GW, Gösele U, Dai S (2007) Chem Mater 19:2383
190. Wang K, Zhang W, Phelan R, Morris MA, Holmes JD (2007) J Am Chem Soc 129:13388
191. Klotz M, Albouy PA, Ayral A, Menager C, Grosso D, Van der Lee A, Cabuil V, Babonneau F, Guizard C (2000) Chem Mater 12:1721
192. Grosso D, Balkenende AR, Albouy PA, Ayral A, Amenitsch H, Babonneau F (2001) Chem Mater 13:1848
193. Yamaguchi A, Uejo F, Yoda T, Uchida T, Tanamura Y, Yamashita T, Teramae N (2004) Nat Mater 3:337
194. Yoo SJ, Ford DM, Shantz DF (2006) Langmuir 22:1839
195. Abetz V, Simon PFW (2005) Block Copolymers I (Advances in Polymer Science), vol 189. Springer, Berlin Heidelberg New York, p 125
196. Bates FS, Fredrickson GH (1990) Annual Rev Phys Chem 41:525
197. Castelletto V, Hamley IW (2004) Curr Opin Solid State Mater Sci 8:426
198. Russell TP (1996) Curr Opin Colloid In 1:107
199. Fasolka MJ, Mayes AM (2001) Annual Rev Mater Res 31:323
200. Chen P, Liang HJ, Shi AC, Yp (2007) Macromolecules 40:7329
201. Li WH, Wickham RA (2006) Macromolecules 39:8492
202. Li WH, Wickham RA, Garbary RA (2006) Macromolecules 39:806
203. Miao B, Yan DD, Wickham RA, Shi AC (2007) Polymer 48:4278
204. Sevink GJA, Zvelindovsky AV, Fraaije J, Huinink HP, Qu (2001) J Chem Phys 115:8226
205. Feng J, Liu HL, Hu Y (2006) Macromol Theor Simul 15:674
206. Xu JB, Wu H, Lu DY, He XF, Zhao YH, Wen H (2006) Molec Simul 32:357
207. He XH, Song M, Liang HJ, Pan CY, Qr (2001) J Chem Phys 114:10510
208. Wang Q (2007) J Chem Phys 126:024903
209. Feng J, Ruckenstein E (2006) Macromolecules 39:4899
210. Chen P, He XH, Liang HJ (2006) J Chem Phys 124:104906
211. Feng J, Ruckenstein E (2006) J Chem Phys 125:164911
212. Yu B, Sun PC, Chen TH, Jin QH, Ding DT, Li BH, Shi AC, Kz (2007) J Chem Phys 127:114906
213. Yu B, Sun PC, Chen TC, Jin QH, Ding DT, Li BH, Shi AC (2006) Phys Rev Lett 96:138306
214. Feng J, Ruckenstein E (2007) J Chem Phys 126:124902
215. Xiao XQ, Huang YM, Liu HL, Hu Y (2007) Macromol Theor Simul 16:166
216. Zhu YT, Jiang W (2007) Macromolecules 40:2872
217. Yu B, Sun PC, Chen TH, Jin QH, Ding DT, Li BH, Shi AC (2007) J Chem Phys 126:204903
218. Erukhimovich I, Johner A (2007) Epl 79:56004
219. Segalman RA, Yokoyama H, Kramer EJ, Jl (2001) Adv Mater 13:1152
220. Cheng JY, Mayes AM, Ross CA, Qa (2004) Nat Mat 3:823
221. Cheng JY, Ross CA, Smith HI, Thomas EL (2006) Adv Mater 18:2505
222. Xiang HQ, Shin K, Kim T, Moon S, McCarthy TJ, Russell TP (2005) J Poly Sci Pol Phys 43:3377

223. Shin K, Xiang HQ, Moon SI, Kim T, McCarthy TJ, Russell TP (2004) Science 306:76
224. Xiang H, Shin K, Kim T, Moon SI, McCarthy TJ, Russell TP (2005) Macromolecules 38:1055
225. Decher G, Hong JD (1991) Makromolekulare Chemie-Macromolecular Symposia 46:321
226. Lvov Y, Decher G, Möhwald H (1993) Langmuir 9:481
227. Decher G (1997) Science 277:1232
228. Feldheim DL, Grabar KC, Natan MJ, Mallouk TE, Vc (1996) J Am Chem Soc 118:7640
229. Schmitt J, Decher G, Dressick WJ, Brandow SL, Geer RE, Shashidhar R, Calvert JM, We (1997) Adv Mater 9:61
230. Caruso F, Lichtenfeld H, Giersig M, Möhwald H (1998) J Am Chem Soc 120:8523
231. Donath E, Sukhorukov GB, Caruso F, Davis SA, Möhwald H (1998) Angew Chem Int Ed Engl 37:2202
232. Caruso F, Caruso RA, Möhwald H (1998) Science 282:1111
233. Jiang CY, Tsukruk VV (2006) Adv Mater 18:829
234. Caruso F (2001) Adv Mater 13:11
235. Li JB, Cui Y (2006) J Nanosci Nanotechnol 6:1552
236. Ai SF, Lu G, He Q, Li JB (2003) J Am Chem Soc 125:11140
237. Liang ZJ, Susha AS, Yu AM, Caruso F (2003) Adv Mater 15:1849
238. Ai SF, He Q, Tao C, Zheng SP, Li JB (2005) Macromol Rapid Commu 26:1965
239. Kim DH, Karan P, Göring P, Leclaire J, Caminade AM, Majoral JP, Gösele U, Steinhart M, Knoll W (2005) Small 1:99
240. Frechet JMJ (1994) Science 263:1710
241. Tomalia DA, Naylor AM, Goddard WA (1990) Angew Chem Int Ed Engl 29:138
242. Bosman AW, Janssen HM, Meijer EW (1999) Chem Rev 99:1665
243. Majoral JP, Caminade AM (1999) Chem Rev 99:845
244. Lu G, Al SF, Li JB (2005) Langmuir 21:1679
245. Lu G, Komatsu T, Tsuchida E (2007) Chem Commun, p 2980
246. Tian Y, He Q, Cui Y, Tao C, Li JB (2006) Chem-Eur J 12:4808
247. Tian Y, He Q, Tao C, Cui Y, Ai S, Li JB (2006) J Nanosci Nanotechnol 6:2072
248. Tian Y, He Q, Tao C, Li JB (2006) Langmuir 22:360
249. Hou SF, Wang JH, Martin CR (2005) Nano Lett 5:231
250. Tian Y, He Q, Cui Y, Li JB (2006) Biomacromolecules 7:2539
251. Hou SF, Wang JH, Martin CR (2005) J Am Chem Soc 127:8586
252. Lee D, Nolte AJ, Kunz AL, Rubner MF, Cohen RE (2006) J Am Chem Soc 128:8521
253. Alem H, Blondeau F, Glinel K, Demoustier-Champagne S, Jonas AM (2007) Macromolecules 40:3366
254. Dai JH, Baker GL, Bruening ML (2006) Analyt Chem 78:135
255. Hou SF, Harrell CC, Trofin L, Kohli P, Martin CR (2004) J Am Chem Soc 126:5674
256. Lee H, Kepley LJ, Hong HG, Mallouk TE, L (1988) J Am Chem Soc 110:618
257. Franzl T, Klar TA, Schietinger S, Rogach AL, Feldmann J, Op (2004) Nano Lett 4:1599
258. Feng CL, Zhong XH, Steinhart M, Caminade AM, Majoral JP, Knoll W (2007) Adv Mater 19:1933

Subject Index

Printing: Krips bv, Meppel, The Netherlands
Binding: Stürtz, Würzburg, Germany